Networks of Networks in Biology
Concepts, Tools and Applications

Biological systems are extremely complex and have emergent properties that cannot be explained or even predicted by studying their individual parts in isolation. The reductionist approach, although successful in the early days of molecular biology, underestimates this complexity. As the amount of available data grows, so it will become increasingly important to be able to analyse and integrate these large data sets. This book introduces novel approaches and solutions to the big data problem in biomedicine, and presents new techniques in the field of graph theory for handling and processing multi-type large data sets. By discussing cutting-edge problems and techniques, researchers from a wide range of fields will gain insights into how to exploit big, heterogeneous data in the life sciences through the concept of 'network of networks'.

Narsis A. Kiani is Assistant Professor and Co-leader of Algorithmic Dynamics in the Department of Oncology-Pathology of the Karolinska Institutet, Sweden. She is passionate about mathematics and is interested in the fundamental question of what observations about effects at the microscopic level can tell us about the macroscopic nature of biological systems and vice versa, and how defects and disorder affect these systems.

David Gomez-Cabrero is the Head of the Translational Bioinformatics Unit at Navarrabiomed, Spain. Since 2009, he has specialized in bioinformatics and data integration analysis, first as a post-doctorate and subsequently as Assistant Professor at the Karolinska Institutet, Sweden, and as Senior Lecturer at King's College London, UK. He collaborates with clinical groups that investigate multiple sclerosis, rheumatoid arthritis, chronic obstructive pulmonary disease (COPD) and cancer, among other diseases.

Ginestra Bianconi is Professor of Applied Mathematics at Queen Mary University of London and a Turing Fellow at The Alan Turing Institute. She is Editor-in-Chief of the *Journal of Physics: Complexity* and Editor of *Scientific Reports* and *PLoS One*. She has published more than 150 articles in network theory and interdisciplinary applications, and is the author of the book *Multilayer Networks: Structure and Function*.

Networks of Networks in Biology

Concepts, Tools and Applications

Edited by

NARSIS A. KIANI
Karolinska Institutet, Stockholm

DAVID GOMEZ-CABRERO
King's College London

GINESTRA BIANCONI
Queen Mary University of London
The Alan Turing Institute

CAMBRIDGE
UNIVERSITY PRESS

CAMBRIDGE
UNIVERSITY PRESS

University Printing House, Cambridge CB2 8BS, United Kingdom

One Liberty Plaza, 20th Floor, New York, NY 10006, USA

477 Williamstown Road, Port Melbourne, VIC 3207, Australia

314321, 3rd Floor, Plot 3, Splendor Forum, Jasola District Centre, New Delhi 110025, India

79 Anson Road, #06–04/06, Singapore 079906

Cambridge University Press is part of the University of Cambridge.

It furthers the University's mission by disseminating knowledge in the pursuit of education, learning, and research at the highest international levels of excellence.

www.cambridge.org
Information on this title: www.cambridge.org/9781108428873
DOI: 10.1017/9781108553711

© Cambridge University Press 2021

First published 2021

Printed in the United Kingdom by TJ Books Limited, Padstow Cornwall

A catalogue record for this publication is available from the British Library.

ISBN 978-1-108-42887-3 Hardback

Contents

Contributors

Rodrigo Bacigalupe VIB-KU Leuven Center for Microbiology, Campus Gasthuisberg, Rega Instituut, Leuven, Belgium

Gordon Ball Unit of Computational Medicine, Department of Medicine Solna, Center for Molecular Medicine, Karolinska Institutet, Stockholm, Sweden (former member)

Ginestra Bianconi School of Mathematical Sciences, Queen Mary University of London, London, UK; The Alan Turing Institute, London, UK

Akram Dehnokhalaji Operations and Information Management Department, Aston Business School, Birmingham, UK

Manlio De Domenico Fondazione Bruno Kessler, Povo, Italy

David Gomez-Cabrero Translational Bioinformatics Unit, Navarrabiomed, Complejo Hospitalario de Navarra (CHN), Universidad Pública de Navarra (UPNA), IdiSNA, Pamplona, Spain; Mucosal and Salivary Biology Division, King's College London Dental Institute, London, UK; Unit of Computational Medicine, Center for Molecular Medicine, Department of Medicine Solna, Karolinska Institutet, Karolinska University Hospital and Science for Life Laboratory, Stockholm, Sweden

Mika Gustafsson Department of Physics, Chemistry and Biology (IFM) Bioinformatics (BION), Linköping University, Linköping, Sweden

Narsis A. Kiani Algorithmic Dynamics Lab, Unit of Computational Medicine, Department of Oncology-Pathology, Centre for Molecular Medicine, Karolinska Institute and SciLifeLab, Stockholm, Sweden; Algorithmic Nature Group, LABORES for the Natural and Digital Sciences, Paris, France

Mikko Kivelä Department of Computer Science, School of Science, Aalto University, Finland

Ingrid Kockum Department of Clinical Neuroscience and Center for Molecular Medicine, Karolinska Institute, Karolinska University Hospital, Stockholm, Sweden

Xabier Martinez de Morentin Translational Bioinformatics Unit, Navarrabiomed, Complejo Hospitalario de Navarra (CHN), Universidad Pública de Navarra (UPNA), IdiSNA, Pamplona, Spain

Jörg Menche CeMM Research Center for Molecular Medicine of the Austrian Academy of Sciences, Vienna, Austria

Nasim Nasrabadi Department of Mathematics, University of Birjand, Iran

Nuria Planell Translational Bioinformatics Unit, Navarrabiomed, Complejo Hospitalario de Navarra (CHN), Universidad Pública de Navarra (UPNA), IdiSNA, Pamplona, Spain

Saeed Shoaie Centre for Host–Microbiome Interactions, Faculty of Dentistry, Oral & Craniofacial Sciences, King's College London, London, UK; Science for Life Laboratory, KTH–Royal Institute of Technology, Stockholm, Sweden

Celine Sin CeMM Research Center for Molecular Medicine of the Austrian Academy of Sciences, Vienna, Austria

Massimo Stella Fondazione Bruno Kessler, Povo, Italy

Jesper Tegnér Biological and Environmental Sciences and Engineering Division, Computer, Electrical and Mathematical Sciences and Engineering Division, King Abdullah University of Science and Technology (KAUST), Thuwal, Saudi Arabia; Unit of Computational Medicine, Center for Molecular Medicine, Department of Medicine Solna, Karolinska Institutet, Karolinska University Hospital and Science for Life Laboratory, Stockholm, Sweden

Hector Zenil Algorithmic Dynamics Lab, Department of Oncology-Pathology, Centre for Molecular Medicine, Karolinska Institute, Stockholm, Sweden; Algorithmic Nature Group, LABORES for the Natural and Digital Sciences, Paris, France

Preface

Network Science is transforming research in different areas of living sciences, ranging from evolutionary biology to medicine. This book aims to introduce the most recent developments of Network Science and its biological applications. In particular, the architecture of biological systems will be characterized using both single networks (monoplexes) and networks of networks (multi-layer networks). Graph theory and social network analysis are classic subjects of mathematics and sociology that have been widely investigated in the twentieth century. However, only since the late 1990s, when several papers on the fundamental design principles of various kinds of large-scale networks were published, was Network Science established as a novel framework to analyse interacting systems, which led to the expansion of the research field of network analysis, with great impact in biology.

After 20 years, this field is still proliferating, and it has become increasingly clear that the challenge of integrating and modelling heterogeneity and complexity of living systems cannot be addressed by using a single-layer network (monoplex).

Instead, the complexity of living systems – such as the interactions between proteins and the microbiome – is captured by a multi-layer network approach. Multi-layer networks are formed by nodes and interactions of different natures, and connotations, forming different networks that interact with each other, creating a network of networks. The investigation of multi-layer networks is a very active research area in Network Science, which in the last ten years has led to the development of several methodologies to extract information for analysing large data sets and integrating the results from different experiments.

With contributions from key leaders, this book discusses topics in Network Science that are increasingly foundational for the quantitative understanding of living systems. Furthermore, it consolidates existing practical and theoretical knowledge of both monoplexes and multi-layer networks. The book balances application and theory to give a unified overview of this interdisciplinary science. It is intended to serve as an introductory text for graduate students and researchers in physics, biology, and biochemistry, and presents ideas and techniques from fields outside the reader's own area of specialization.

Aims of this Book

This book focuses on network-inspired approaches for the analysis and integration of the large data sets currently prevalent in life sciences. The principal aim of this work is to give a comprehensive overview of new techniques in the field of graph theory for handling and processing multi-type large data sets. By discussing state-of-the-art problems and techniques, this book offers researchers from a wide range of areas the unique opportunity to gain insights for exploiting the richness of life sciences data sets through the concept of 'networks of networks'. It presents a timely, multi-authored compendium representing a diverse set of backgrounds and methods developed in this area. Contributions have been selected and compiled to introduce the concept and cover different methods in a way that is accessible to people from diverse backgrounds. We hope that researchers from different areas of network analysis will learn new aspects and future directions of this emerging field.

How to Read this Book

In order to reach a broad spectrum of readers – biologists, biochemists, computer scientists, bioinformaticians – the book does not require a deep knowledge of computer science or biology. Instead, the reader will learn about graph theory, graph algorithms, and network analysis, as well as biology.

This book consists of five parts: Part I provides a brief overview of biological networks and graph theory; Part II includes chapters discussing network analysis; Part III presents chapters introducing multiplex and multi-layer networks and tools to extract relevant information from these networks of networks; Part 4 includes chapters presenting the application to real biological case studies and the initial insights gained in this relatively new field of research; finally, Part V provides concluding remarks and an outline of the future directions of the field. The book builds upon the introductory chapters and ends by addressing emerging trends in this growing, vibrant research area. Each chapter can be studied independently.

PART I
NETWORKS IN BIOLOGY

1 An Introduction to Biological Networks

Nuria Planell, Xabier Martinez de Morentin and David Gomez-Cabrero

1.1 Biology Needs to be Analysed Like a System

From basic biology to clinical research, scientists are trying to elucidate the mechanisms underlying the regulation of cells in order to understand their origin, evolution and behaviour in health and disease. Nowadays, we know that the human body comprises a considerable number of different cell types working in coordination. Within a cell, the following framework depicts our understanding of *the biological information flow from the genome to the phenome*: first, the DNA molecules (genomics) are transcribed to mRNA (transcriptomics) and then translated into proteins (proteomics), which can catalyse reactions that act on and give rise to metabolites (metabolomics), glycoproteins and oligosaccharides (glycomics), and various lipids (lipidomics). Finally, these proteins and biomolecules are involved in different metabolic pathways and cellular processes that, in conjunction, dictate the cell behaviour or phenotype [1].

The study of each one of these layers of information (genomics, transcriptomics and proteomics, among others) independently has been extensive and, as a result, there is significant knowledge of the sophisticated machinery that orchestrates the cellular processes. Furthermore, within each layer, many single features (e.g. single genes) have been the target of extensive research, such as the TP53 protein [2–4]. The single-feature analysis derives *partially* from historical technical limitations and from the *belief* that one gene produced a single protein and that one protein had a single function. As a result, there are many *single-gene vs single disease analyses* [2–4]. However, many genes produce several protein isoforms and proteins may have different functions and cellular roles, depending on their environment [5]. Most importantly, many features interact and the 'single-feature' analysis does not allow characterizing such interactions or the behaviours derived from them. Importantly, most cellular functions are organized as highly connected sets of genes and/or proteins and/or metabolites communicating through biochemical and physical interactions. Therefore, *biology needs to move to a holistic view and start to explore all the biological information in an integrated way*: **as a system**. Now, we need *to identify (the best) ways to model biological systems* [6, 7].

One way is to focus on the features and their interactions (whatever the nature of such interactions) and, as a result, a biological system can be depicted as a *network* [8]. In such a *biological* network, the components (nodes) can be genes, proteins or metabolites, among other elements, and the interactions can be physical interactions, biochemical interactions or co-expression, among others.

To illustrate the concept, we will detail an example: a pathogen (for instance, a virulent strain of *Escherichia coli*) infecting our body. When this happens, the immunological response is activated to eradicate the infection and restore a healthy status. At the cellular level, it means that different processes are initiated to produce a pathogen-related response. As a brief description, these processes start with a signal (stimulus) that triggers a sequence of (chemical or physical) signals that are transmitted through the cell, provoking a signal cascade that results in a cellular response. Any process that starts from a particular stimulus and is transformed into a biochemical signal throughout the cell is known as a *signal transduction process* (and these are all good candidates for network modelling).

As a detailed example, we consider one of the signal transduction processes activated as a pathogen-related response, the TLR4 (Toll-like receptor 4) signal transduction pathway. The interaction between the pathogenic molecule and the cellular receptor TLR4 initiates the signal transduction by recruiting intracellular adaptor molecules such as myeloid differentiation factor 88 (MyD88) and TIRF-related adaptor proteins. Depending on the adaptor proteins recruited, two different signal cascades can take place: one that depends on the MyD88 molecule and another which is TRIF-dependent. Following the MyD88-dependent pathway, after the recruitment of adaptor proteins, TNF receptor-associated factor 6 (TRAF6) is activated to interact with the second complex of proteins (TAK1 and TAB2/3). Going forward, mitogen protein kinases (MAPKs; MKK3/6 and MKK4/7) and another complex of proteins (NEMO/IKK complex) are activated, leading to the activation of AP1 (through p38 or c-Jun N-terminal kinase (JNK)) and NF-κB, respectively; all are involved in the transcription control of pro-inflammatory cytokines (IL-6, IL-12, TNF-α, etc.). The MyD88-independent pathway recruits TRIF-dependent adaptor proteins and starts the signal cascade by binding to the IKK-related kinase TBK1 and IKKε, which mediates direct phosphorylation of IRF3 transcription factor. IRF3 will migrate to the cellular nucleus and promote the transcription of IFN-inducible genes [9, 10]. Briefly, from the initial pathogenic stimulus, a signal cascade starts to lead to the production of inflammatory-related cytokines.

In Figure 1.1a (inspired and partially adapted from [10]), the infection process described is depicted, where the proteins or protein complexes are the nodes and the physical or biochemical interactions the edges. Such description can be further summarized into a network, as shown in Figure 1.1b, where *elements* of the information are ignored (e.g. location in the cell or the type of interaction) and only proteins (*nodes*) and interactions (*edges*) are kept. Following both representations, we can identify and follow the signal cascade from the initial stimulus to the final cellular response. The network representation has a mathematical description and notation that will be introduced in the next section (and further discussed in Chapter 2). Finally, in Figure 1.1c we observe that the network can also be stored as a matrix, where rows and columns denote the proteins, and for an entity in the matrix a '1' (dark grey in the figure) denotes an interaction between both proteins.

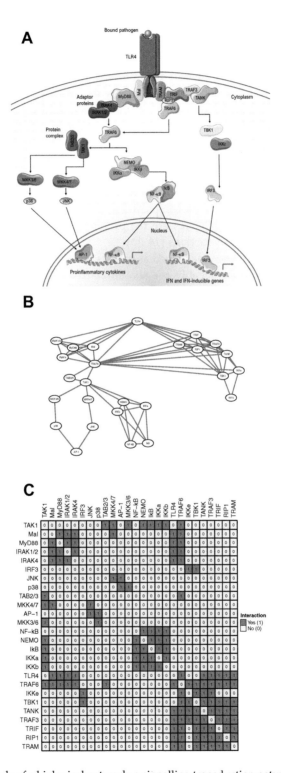

Figure 1.1   Example of a biological network: a signalling transduction network.
(a) Biological description. (b) Network description of (a). (c) Contingency matrix description
of (a). See text for details of the biological network described [10].

1.2   Introduction to Networks

In the previous section, we established that biological systems can be *modelled* as networks. The role of the modelling here is to provide a holistic description of a system (derived from the biological information) in a way that allows studying characteristics of the systems that cannot be derived from the collection of per-feature characteristics ('stamp collection' [7]). In biological networks, *nodes* can represent any type of biological molecule or even a complex of molecules. *Edges* can represent any type of relationship between a pair of nodes; for example, edges may represent that two molecules are present in the same tissue, are related to the same disease, are part of the same biological process [11] or the same molecular function [12], or similar expression levels [13], among other relationships. In Figure 1.1, a protein–protein interaction network is depicted in which nodes are the proteins and, in some cases, the edges represent known physical interactions [14, 15].

Biological networks can be described as graphs; and, while in the text we will use graph and network interchangeably, we should clarify that network analysis is the study of graphs when they represent relations (symmetric or asymmetric) between discrete objects [16, 17]. Interestingly, the concept of graph theory was initially developed as a tool to solve *mathematical riddles*. The first (and most famous) riddle is the problem of the bridges of Königsberg: the town Königsberg had seven bridges and the problem was to visit all parts of the city while crossing each bridge only once [18]. Euler proved in 1736 that there was no feasible solution [19].

Importantly, around the end of the 1950s, the analysis and generation of random graphs was proposed by Erdös and Rényi [20] and simultaneously by Gilbert [21]. A random graph studies the uniformly random selection of graphs from the set of all possible graphs with N nodes and M edges, with N and M being arbitrary numbers. Interestingly, it was observed that those models were not able to capture a property observed in most 'real-life' networks: small-world properties. A significant small-world property is the short average path length necessary to connect every pair of nodes. Watts and Strogatz proposed a model to generate small-world random graphs [22]. However, those graphs did not generate another 'real-life' network property: 'hubs'. Hubs are (a small number of) nodes with a more extensive than average number of edges to/from other nodes; the property is termed 'scale-free'. Barabási and Albert studied scale-free graph properties [23].

The analysis of random vs non-random graphs is of particular relevance, which we will explore further in a later section, because in biological systems (as well as observed in social networks) the graphs associated are not random graphs as defined by Erdös and Rényi [20]. For instance, there are nodes with an increased number of edges. In gene networks, these nodes are known as 'hubs' or 'master regulators', and they are of interest because they may show an association with specific biological processes.

Computationally, the process of drawing biological systems into networks can be described mathematically by adjacency matrices (see Figure 1.1c). In such a matrix, both columns and edges are the nodes, and every position (node$_i$,node$_j$) may denote the existence/non-existence of an edge as a binary 1/0 (e.g. a protein–protein interaction [24]), or they may specify numerical 'weights' that may be associated with the strength of the relationships. Those weights could be computed as a measure of similarity between the nodes using, for instance, correlation or mutual information

[25], among other measures. It is essential to specify that the selection of the distance measure may shape a biological network differently [26].

1.3 Types of Biological Networks

As previously presented, a biological system can be represented as a biological network, and it may include several groups of coordinated subsystems. From the molecular level up to a whole biological system, one can think in different types of networks: molecular networks, cell-to-cell networks, host–microbiome networks and systems medicine networks. Moreover, each one of these networks can be divided into different subsystems.

Within molecular networks, the most relevant ones are protein–protein interaction (PPI) networks, gene regulatory networks, signal transduction networks, metabolomics or biochemical networks and functional or co-expression networks. The nodes of these networks are genes and/or proteins and/or metabolites, and the edges are physical or biochemical interactions, co-expression patterns, etc. [17, 27]. A schematic representation of the different types of molecular networks is shown in Figure 1.2.

Protein–protein interaction networks are fundamental in biological functions. Protein interactions determine molecular and cellular mechanisms that control healthy and diseased states in organisms. In these networks, nodes represent proteins and edges represent a physical interaction between two proteins. The edges are non-directed, as it cannot be said which protein binds the other; that is, which partner functionally influences the other. Within the example described in Figure 1.1a, several PPI networks can be defined. The interaction between different adaptor molecules and the TLR4 gives a complex structure that is *per se* a PPI.

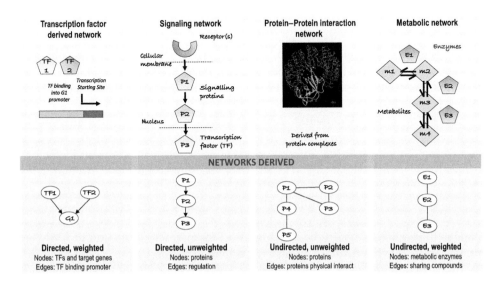

Figure 1.2 Examples of types of biological networks. The image from the Protein-Protein interaction section was created in 2002 by Dcrjsr, and is licensed under the Creative Commons Attribution 3.0 Unported licence.

The structures and dynamics of protein networks are disturbed in complex diseases such as cancer [24] and autoimmune disorders. Therefore, such networks facilitate the understanding of these mechanisms in both pathogenic or physiologic scenarios and can be translated into effective diagnostic and therapeutic strategies [28].

To generate PPI networks, besides the various experimental methods, a variety of large biological databases that collect and organize PPI information are available, most of which are organism-specific. Among them are the Yeast Proteome Database (YPD) [29], the Munich Information Center for Protein Sequences (MIPS) [30], the Molecular Interactions (MINT) database [31], the IntAct database [32], the Database of Interacting Proteins (DIP) [33], the Biomolecular Interaction Network Database (BIND) [34], the BioGRID database [35], the Human Protein Reference Database (HPRD) [36], the HPID [37] and the DroID for *Drosophila* [38]. Additionally, well-documented services based on text-mining analysis provide relevant resources, including the Stitch and String databases [39, 40].

Gene regulatory networks give information concerning the control of gene expression in cells. Nodes are either a transcription factor or a putative DNA regulatory element, and directed edges represent the physical binding of transcription factors to such regulatory elements. Edges are directed: *incoming* (transcription factor binds a regulatory DNA element) or *outgoing* (regulatory DNA element bound by a transcription factor). In addition to transcription factor activities, overall gene transcript levels are also regulated post-transcriptionally by microRNAs (miRNAs), short noncoding RNAs that bind to complementary cis-regulatory RNA sequences usually located in 30 untranslated regions (UTRs) of target mRNAs. Then, edges can also be denoted as *incoming* (miRNA binds a 30UTR element) or *outgoing* (30UTR element bound by an miRNA).

These networks use a directed graph representation to model the way proteins and other biological molecules are involved in gene expression, and they aim to describe the order of the events that take place in different stages of the process. Following the example in Figure 1.1a, the associated (not in the figure) regulatory network of the activated transcription factors AP1 and NF-κB could be detailed.

To generate this regulatory networks, protein–DNA interaction data is collected in databases like JASPAR [41], TRANSFAC [42] or B-cell Interactome (BCI) [43], while post-translational modification can be found in databases like Phospho.ELM [44], NetPhorest [45] or PHOSIDA [46].

Signal transduction networks connect receptors and many different cellular machines. Such networks not only receive and transmit signals, but also process information. To represent the series of interactions between the different biological entities (nodes) such as proteins, chemicals or macromolecules and to investigate how signal transmission is performed either from the outside to the inside of the cell or within the cell, multi-edged directed graphs are used. One example of these signal cascades is shown in Figure 1.1a. Given a pathogenic stimulus, a signal is transmitted through the cell to give a response. Depending on the cellular circumstances (environmental parameters), different responses can be triggered; in that way, the environment could trigger for a MyD88-dependent or -independent response in the case of the TLR4 signalling pathway. Some sources of information regarding signal transduction pathways are the MiST [47] and TRANSPATH [48] databases.

Metabolomics or biochemical networks describe a series of chemical reactions occurring within a cell at different time points. The enzymes play the primary role within a metabolic network since they are the main determinants in catalysing biochemical reactions. Often, enzymes are dependent on other cofactors, such as vitamins, for proper functioning.

In graph representation of metabolic networks, nodes are metabolites and edges are either the enzymes that catalyse these reactions or the reactions that convert one metabolite into another. Edges can be directed or undirected, depending on the reversibility of a given reaction. Among the several databases holding information about biochemical networks, some of the most popular are the Kyoto Encyclopedia of Genes and Genomes (KEGG) [49], EcoCyc [50], BioCyc [51] and metaTIGER [52].

Functional networks are gene co-expression networks. The reasoning used to define this type of network is that associated proteins are more likely to be encoded by genes with similar transcription profiles [53, 54]. In these networks, nodes represent genes and edges link pairs of genes that show correlated co-expression above a set threshold based on an association measure such as the Pearson correlation coefficient or mutual information [55]. In the example shown in Figure 1.1a, the set of genes whose transcription is regulated by NF-κB and AP1, such as IL-6, IL-12, IL-1 and TNF-α, may show statistically significant correlation because they are involved in the same biological process [56].

Beyond molecules, **cell–cell communication (CCC) networks** can also be defined. This kinds of networks describe the cross-talk between cells. In those networks, nodes are different cell types and the edges are receptor–ligand interactions. A CCC network is a directional bipartite graph that is usually constructed based on the differential over-expression of ligand and receptor genes of the cell types of interest [57].

Given the complex system that defines a whole organism and the functional interdependencies between the molecular components shown in a human cell, we observe that most diseases are rarely a consequence of an abnormality in a single gene. Instead, the disease phenotype reflects the perturbations of a complex intracellular network. The identification of these perturbed networks defined as disease modules can allow the identification of molecular relationships between apparently distinct pathologic phenotypes. These disease connections can be presented as a **disease network**, where nodes are disease and diseases are connected if they share one or several disease-associated genes or if they are both associated with enzymes that catalyse adjacent reactions. In metabolic diseases, links induced by shared metabolic pathways are expected to be more relevant than links based on shared genes [58]. To construct this kind of network, available resources are the gene–disease associations collected in the OMIM [59], KEGG [60] and BiGG [61] database.

Other approaches are emerging within systems medicine, including **drug–target networks** and **drug–drug networks**. Both drug–target and drug–drug networks will help in new drug development as they are implicated in drug discovery and prediction of adverse effects [62, 63]. Those types of networks are also described in Chapter 9.

Finally, **microbiome–host networks** can also be defined. The role of the microbiome in human health and disease has received greater interest during recent years as the microbiome is involved in metabolism, physiology, nutrition and different immunological functions. For more in-depth information on microbiome and host–microbiome networks, see Chapter 11.

In summary, several types of networks can be defined in biology in order to explain and simplify complex systems. However, these approaches are restricted to the amount of information known; a vast amount of interactions is thought to be unknown. Consequently, biological networks should be considered as a dynamic field that will evolve over time, depending on knowledge generation and curation.

1.4 Mathematical Properties of Biological Networks

In biological networks, as well as in social networks, the distribution of the number of edges incident upon a node – denoted as *node degree centrality* measure – follows a power law distribution, $P(k) = Ak^{2-y}$ [64], that is not observed in random graphs. As a result of the power-law distribution there will be high diversity of node degrees; this characteristic is known as **scale-free** [23]. A second property is the **small world** [22], which denotes that the 'shortest path' (or the collection of nodes) needed to communicate a pair of nodes is reduced compared to random networks.

An additional property of interest in networks is connectivity, which estimates (and identifies) the minimum number of edges (or nodes) required to separate nodes into isolated subgraphs. Isolated subgraphs are groups of nodes that cannot describe a path connecting them. In Figure 1.1a, the elimination of edges (p38,AP-1), (JNK,AP-1) and (NF-κKB,NF-κB) would generate two subgraphs.

There are also measures of interest that define the relevance of a node, such as centrality measurements. Beyond node degree centrality, *betweenness centrality* quantifies the number of times a node appears in shortest paths between pairs of nodes or *closeness centrality* quantifies the average length of the paths between the node of interest and any other node, among other measures [8]. These properties are described and discussed in Chapter 3.

1.5 Storing and Visualizing Networks

Networks are a useful tool for modelling and studying most biological systems. While the mathematical tools for their analysis are relevant, the storage and visualization of networks are also relevant because they provide powerful exploratory tools.

For storing and communication, the Systems Biology Markup Language (SBML) [65] provides a representation format based on XML, which allows the communication and storage of computational models of biological processes. It's an open-source framework and nowadays is the standard for representing computational models in systems biology.

For visualization, there are many tools available, among the most popular being Cytoscape [66] and Gephi [67]. Both tools provide methods for visualization, but also network analysis (including the estimation of centrality measures) or interfaces with programming languages such as R. Importantly, network visualization is a complex problem by itself, because it requires describing in two dimensions a set of features and their connections. There are several methodologies available for the projection of networks in two dimensions (named layout). Several examples of the network shown in Figure 1.1 are shown in Figure 1.3.

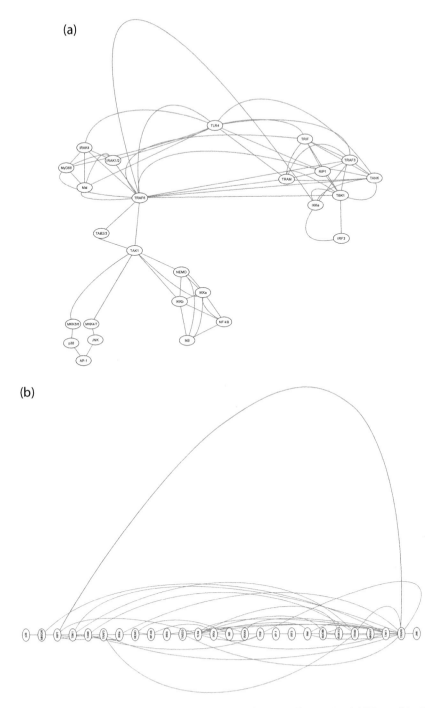

Figure 1.3 Two networks layouts from the network shown in Figure 1.1. (a) Hierarchical layout. (b) Stacked node layout.

Finally, is it also relevant to mention the existence of standards for annotation in network visualizations; a good example is Systems Biology Graphical Notation (SBGN) [68, 69].

1.6   Conclusions and 'Networks of Networks in Biology'

Two major conclusions are delivered from the current discussion. The first is that *to understand any biological system, it is necessary to shift from a collection of facts towards a truly holistic approach that considers all elements of a system*. To this end, and as a second conclusion, *network analysis provides a modelling framework that summarizes a given system based on its elements and their interactions*. In Figure 1.1 we showed an example of such applications, and in Figure 1.2 we have briefly summarized current network-based applications in systems biology. Additionally, we have shown that over the past two decades, there has been extensive development of tools for generating, visualizing, analysing, storing and sharing networks. Importantly, while in the current chapter we have addressed the *intuitive* concept of networks, in the coming chapters the mathematical description of networks will be provided in detail (see Chapter 2).

A final note is that to advance the holistic approach, all the different layers described in Figure 1.2 need to be integrated. Classical network analysis did not consider *in detail* such types of networks in which multiple types of nodes may be considered, where pairs of nodes may be associated by several edges that depict different types of connections (e.g. 'transcription-factor to gene' and 'correlation between genes'). Therefore, new types of networks and their analytical tools are required. This book discusses multi-layer networks as the next set of network analysis tools [70, 71].

References

[1] Joyce AR, Palsson BØ. The model organism as a system: integrating 'omics' data sets. *Nat Rev Mol Cell Biol*. 2006;7:198–210. doi:10.1038/nrm1857

[2] Cheng Y-K, Beroukhim R, Levine RL, et al. A mathematical methodology for determining the temporal order of pathway alterations arising during gliomagenesis. *PLoS Comput Biol*. 2012;8:e1002337. doi:10.1371/journal.pcbi.1002337

[3] Soussi T. The TP53 gene network in a postgenomic era. *Hum Mutat*. 2014;35:641–642. doi:10.1002/humu.22562

[4] Soussi T, Wiman KG. TP53: an oncogene in disguise. *Cell Death Differ*. 2015;22:1239–1249. doi:10.1038/cdd.2015.53

[5] Horowitz NH, Bonner D, Mitchell HK, Tatum EL, Beadle GW. Genic Control of biochemical reactions in neurospora. *Am Nat*. 1945;79:304–317. doi:10.1086/281267

[6] Kitano H. Systems biology: a brief overview. *Science*. 2002;295:1662–1664. doi:10.1126/science.1069492

[7] Gomez-Cabrero D, Tegnér J. Iterative systems biology for medicine: time for advancing from network signatures to mechanistic equations. *Curr Opin Syst Biol*. 2017;3. doi:10.1016/j.coisb.2017.05.001

[8] Jeong H, Tombor B, Albert R, Oltvai ZN, Barabàsi A-L. The large-scale organization of metabolic networks. *Nature*. 2000;407:651–654. doi:10.1038/35036627

[9] Mukherjee S, Karmakar S, Babu SPS. TLR2 and TLR4 mediated host immune responses in major infectious diseases: a review. *Brazilian J Infect Dis*. 2016;20:193–204. doi:10.1016/j.bjid.2015.10.011

[10] Mogensen TH. Pathogen recognition and inflammatory signaling in innate immune defenses. *Clin Microbiol Rev*. 2009;22:240–273. doi:10.1128/CMR.00046-08

[11] Blake JA, Christie KR, Dolan ME, et al. Gene Ontology Consortium: going forward. *Nucleic Acids Res*. 2015;43: D1049–D1056. doi:10.1093/nar/gku1179

[12] Ashburner M, Ball CA, Blake JA, et al. Gene ontology? Tool for the unification of biology. *Nat Genet*. 2000;25:25–29. doi:10.1038/75556.Gene

[13] Hong S, Chen X, Jin L, Xiong M. Canonical correlation analysis for RNA-seq co-expression networks. *Nucleic Acids Res*. 2013;41:1–15. doi:10.1093/nar/gkt145

[14] Luck K, Sheynkman GM, Zhang I, Vidal M. Proteome-scale human interactomics. *Trends Biochem Sci*. 2017;42:342–354. doi:10.1016/j.tibs.2017.02.006

[15] Rolland T, Taan M, Charloteaux B, et al. A proteome-scale map of the human interactome Network. *Cell*. 2014;159:1212–1226. doi:10.1016/j.cell.2014.10.050

[16] Koutrouli M, Karatzas E, Paez-Espino D, Pavlopoulos Georgios A. A guide to conquer the biological network era using graph theory. *Front Bioeng Biotechnol*. 2020;8. doi:10.3389/fbioe.2020.00034.

[17] Pavlopoulos GA, Secrier M, Moschopoulos CN, et al. Using graph theory to analyze biological networks. *BioData Min*; 2011;4:10. doi:10.1186/1756-0381-4-10

[18] Shields R. Cultural topology: the seven bridges of Königsburg, 1736. *Theor Cult Soc*. 2012;29:43–57. doi:10.1177/0263276412451161

[19] Euler L. Solutio problematis ad geometriam situs pertinentis. *Comment Acad Sci U Petrop*. 1736;8:128–140.

[20] Erdös P, Rényi A. On random graphs I. *Publ Math Debrecen*. 1959;6:290.

[21] Gilbert EN. Random graphs. *Ann Math Stat*. 1959;30:1141–1144.

[22] Watts DJ, Strogatz SH. Collective dynamics of 'small-world' networks. *Nature*. 1998;393:440–442.

[23] Barabási A-L. Scale-free networks: a decade and beyond. *Science*. 2009;325:412–3. doi:10.1126/science.1173299

[24] Lievens S, Van der Heyden J, Masschaele D, et al. Proteome-scale binary interactomics in human cells. *Mol Cell Proteomics*. 2016;15:3624–3639. doi:10.1074/mcp.M116.061994

[25] Meyer PE, Lafitte F, Bontempi G. minet: a R / bioconductor package for inferring large transcriptional networks using mutual information. *BMC Bioinformat* 2008;10:1–10. doi:10.1186/1471-2105-9-461

[26] Albert R. Network inference, analysis, and modeling in systems biology. *Plant Cell*. 2007;19:3327–3338. doi:10.1105/tpc.107.054700

[27] Vidal M, Cusick ME, Barabási A-L. Interactome networks and human disease. *Cell*. 2011;144:986–998. doi:10.1016/j.cell.2011.02.016

[28] Safari-Alighiarloo N, Taghizadeh M, Rezaei-Tavirani M, Goliaei B, Peyvandi AA. Protein–protein interaction networks (PPI) and complex diseases. *Gastroenterol Hepatol*. 2014;7:17–31.

[29] Hodges PE, Payne WE, Garrels JI. The Yeast Protein Database (YPD): a curated proteome database for *Saccharomyces cerevisiae*. *Nucleic Acids Res*. 1998;26:68–72.

[30] Mewes HW, Amid C, Arnold R, et al. MIPS: analysis and annotation of proteins from whole genomes. *Nucleic Acids Res*. 2004;32:41D–44. doi:10.1093/nar/gkh092

[31] Zanzoni A, Montecchi-Palazzi L, Quondam M, et al. MINT: a Molecular INTeraction database. *FEBS Lett*. 2002;513:135–140.

[32] Kerrien S, Alam-Faruque Y, Aranda B, et al. IntAct: open source resource for molecular interaction data. *Nucleic Acids Res*. 2007;35: D561–D565. doi:10.1093/nar/gkl958

[33] Xenarios I, Rice DW, Salwinski L, et al. DIP: the database of interacting proteins. *Nucleic Acids Res*. 2000;28:289–291.

[34] Bader GD, Betel D, Hogue CWV. BIND: the Biomolecular Interaction Network Database. *Nucleic Acids Res*. 2003;31:248–250.

[35] Stark C, Breitkreutz B-J, Reguly T, et al. BioGRID: a general repository for interaction datasets. *Nucleic Acids Res*. 2006;34: D535–D539. doi:10.1093/nar/gkj109

[36] Goel R, Harsha HC, Pandey A, Prasad TSK. Human Protein Reference Database and Human Proteinpedia as resources for phosphoproteome analysis. *Mol Biosyst*. 2012;8:453–463. doi:10.1039/c1mb05340j

[37] Han K, Park B, Kim H, Hong J, Park J. HPID: the Human Protein Interaction Database. *Bioinformatics*. 2004;20:2466–2470. doi:10.1093/bioinformatics/bth253

[38] Yu J, Pacifico S, Liu G, Finley Jr. RL. DroID: the Drosophila Interactions Database, a comprehensive resource for annotated gene and protein interactions. *BMC Genomics*. 2008;9:461. doi:10.1186/1471-2164-9-461

[39] Kuhn M, Szklarczyk D, Franceschini A, et al. STITCH 2: an interaction network database for small molecules and proteins. *Nucleic Acids Res*. 2010;38: D552–D556. doi:10.1093/nar/gkp937

[40] Jensen LJ, Kuhn M, Stark M, et al. STRING 8: a global view on proteins and their functional interactions in 630 organisms. *Nucleic Acids Res*. 2009;37: D412–D416. doi:10.1093/nar/gkn760

[41] Sandelin A, Alkema W, Engström P, Wasserman WW, Lenhard B. JASPAR: an open-access database for eukaryotic transcription factor binding profiles. *Nucleic Acids Res.* 2004;32: D91–D94. doi:10.1093/nar/gkh012

[42] Matys V, Fricke E, Geffers R, et al. TRANSFAC: transcriptional regulation, from patterns to profiles. *Nucleic Acids Res.* 2003;31:374–378.

[43] Lefebvre C, Lim WK, Basso K, Favera RD, Califano A. A context-specific network of protein–dna and protein–protein interactions reveals new regulatory motifs in human B cells. In: *Systems Biology and Computational Proteomics.* Berlin: Springer, 2007, pp. 42–56.

[44] Dinkel H, Chica C, Via A, et al. Phospho.ELM: a database of phosphorylation sites – update 2011. *Nucleic Acids Res.* 2011;39: D261–D267. doi:10.1093/nar/gkq1104

[45] Miller ML, Jensen LJ, Diella F, et al. Linear motif atlas for phosphorylation-dependent signaling. *Sci Signal.* 2008;1:ra2. doi:10.1126/scisignal.1159433

[46] Gnad F, Ren S, Cox J, et al. PHOSIDA (phosphorylation site database): management, structural and evolutionary investigation, and prediction of phosphosites. *Genome Biol.* 2007;8:R250. doi:10.1186/gb-2007-8-11-r250

[47] Ulrich LE, Zhulin IB. MiST: a microbial signal transduction database. *Nucleic Acids Res.* 2007;35:D386–D390. doi:10.1093/nar/gkl932

[48] Schacherer F, Choi C, Götze U, et al. The TRANSPATH signal transduction database: a knowledge base on signal transduction networks. *Bioinformatics.* 2001;17:1053–1057.

[49] Kanehisa M, Goto S, Furumichi M, Tanabe M, Hirakawa M. KEGG for representation and analysis of molecular networks involving diseases and drugs. *Nucleic Acids Res.* 2010;38: D355–D360. doi:10.1093/nar/gkp896

[50] Keseler IM, Bonavides-Martinez C, Collado-Vides J, et al. EcoCyc: a comprehensive view of *Escherichia coli* biology. *Nucleic Acids Res.* 2009;37:D464–D470. doi:10.1093/nar/gkn751

[51] Karp PD, Ouzounis CA, Moore-Kochlacs C, et al. Expansion of the BioCyc collection of pathway/genome databases to 160 genomes. *Nucleic Acids Res.* 2005;33:6083–6089. doi:10.1093/nar/gki892

[52] Whitaker JW, Letunic I, McConkey GA, Westhead DR. metaTIGER: a metabolic evolution resource. *Nucleic Acids Res.* 2009;37:D531–D538. doi:10.1093/nar/gkn826

[53] Kemmeren P, van Berkum NL, Vilo J, et al. Protein interaction verification and functional annotation by integrated analysis of genome-scale data. *Mol Cell.* 2002;9:1133–1143.

[54] Ge H, Liu Z, Church GM, Vidal M. Correlation between transcriptome and interactome mapping data from *Saccharomyces cerevisiae*. *Nat Genet.* 2001;29:482–486. doi:10.1038/ng776

[55] Margolin AA, Nemenman I, Basso K, et al. ARACNE: an algorithm for the reconstruction of gene regulatory networks in a mammalian cellular context. *BMC Bioinformatics*. 2006;7(Suppl 1):S7. doi:10.1186/1471-2105-7-S1-S7

[56] Oeckinghaus A, Ghosh S. The NF-kappaB family of transcription factors and its regulation. *Cold Spring Harb Perspect Biol*. 2009;1:a000034. doi:10.1101/cshperspect.a000034

[57] Qiao W, Wang W, Laurenti E, et al. Intercellular network structure and regulatory motifs in the human hematopoietic system. *Mol Syst Biol*. 2014;10:741. doi:10.15252/MSB.20145141

[58] Barabási A-L, Gulbahce N, Loscalzo J. Network medicine: a network-based approach to human disease. *Nat Rev Genet*.2011;12:56–68. doi:10.1038/nrg2918

[59] Hamosh A, Scott AF, Amberger JS, Bocchini CA, McKusick VA. Online Mendelian Inheritance in Man (OMIM), a knowledgebase of human genes and genetic disorders. *Nucleic Acids Res*.2005;33:D514–D517. doi:10.1093/nar/gki033

[60] Kanehisa M, Goto S. KEGG: Kyoto encyclopedia of genes and genomes. *Nucleic Acids Res*. 2000;28:27–30.

[61] Schellenberger J, Park JO, Conrad TM, Palsson BØ. BiGG: a Biochemical Genetic and Genomic knowledgebase of large scale metabolic reconstructions. *BMC Bioinformatics*. 2010;11:213. doi:10.1186/1471-2105-11-213

[62] Takeda T, Hao M, Cheng T, Bryant SH, Wang Y. Predicting drug–drug interactions through drug structural similarities and interaction networks incorporating pharmacokinetics and pharmacodynamics knowledge. *J Cheminform*. 2017;9:16. doi:10.1186/s13321-017-0200-8

[63] Luo Y, Zhao X, Zhou J, et al. A network integration approach for drug–target interaction prediction and computational drug repositioning from heterogeneous information. *Nat Commun*. 2017;8:573. doi:10.1038/s41467-017-00680-8

[64] Albert R. Scale-free networks in cell biology. *J Cell Sci*. 2005;118:4947–4957. doi:10.1242/jcs.02714

[65] Krause F, Uhlendorf J, Lubitz T, et al. Annotation and merging of SBML models with semantic SBML. *Bioinformatics*. 2010;26:421–422. doi:10.1093/bioinformatics/btp642

[66] Cline MS, Smoot M, Cerami E, et al. Integration of biological networks and gene expression data using Cytoscape. *Nat Protoc*. 2007;2:2366–2382. doi:10.1038/nprot.2007.324

[67] Bastian M, Heymann S, Jacomy M. Gephi: an open source software for exploring and manipulating networks [Internet]. 2009. Available at: www.aaai .org/ocs/index.php/ICWSM/09/paper/view/154

[68] Novère N Le, Hucka M, Mi H, et al. The Systems Biology Graphical Notation. *Nat Biotechnol*. 2009;27:735–741. doi:10.1038/nbt.1558

[69] Czauderna T, Klukas C, Schreiber F. Editing, validating and translating of SBGN maps. *Bioinformatics*. 2010;26:2340–2341. doi:10.1093/bioinformatics/btq407

[70] Mangioni G, Jurman G, DeDomenico M. Multilayer flows in molecular networks identify biological modules in the human proteome. *IEEE Trans Netw Sci Eng.* doi:10.1109/TNSE.2018.2871726

[71] De Domenico M, Solé-Ribalta A, Cozzo E, et al. Mathematical formulation of multilayer networks. *Phys Rev X.* 2014;3:1–15. doi:10.1103/PhysRevX.3.041022

2 Graph Theory

Akram Dehnokhalaji and Nasim Nasrabadi

2.1 Introduction

The concept of graphs was first introduced by Leonhard Euler for the famous historical problem 'The Seven Bridges of Königsberg'. The city of Königsberg in Prussia (now Kaliningrad, Russia) was set on both sides of the Pregel river and included two large islands and two mainland regions. The four districts were connected to each other by seven bridges. The citizens of Königsberg used to spend Sunday afternoons walking around their beautiful city and decided to create a game for themselves. The question was whether it is possible to walk through all four parts of the city while crossing each of the seven bridges only once, and return to the starting point. Although none of the citizens could invent such a route, it seemed impossible to prove that such a path did not exist. Euler believed this problem was related to something Leibniz referred to as 'geometriasitus', or geometry of position. This so-called geometry of position is what is now called graph theory, which Euler introduced and utilized to solve this famous problem. In 1736, Euler published his first paper on this problem. He proved that there does not exist a route by which one can visit all four regions of the city while crossing each of the seven bridges only once. There are many other real-life applications with the same characteristics that can be studied using graph theory.

In mathematics, and more specifically in graph theory, a graph is a structure made of a set of objects – namely vertices, nodes or points – in which some pairs of the objects are in some sense 'related', where each of the related pairs of vertices is called an edge, arc or line. In a diagram, a graph is depicted as a set of dots (for vertices – joined by lines or curves (for edges). Many practical problems can be represented by graphs. For instance, several kinds of relations and processes in physical, biological, social and information systems can be modelled by graphs. It should be clarified that the terms 'graph' and 'network' are used interchangeably in the literature. This chapter provides a summary of basic definitions and concepts in graph theory.

2.2   Basic Concepts and Definitions

A graph $G = (V, E)$ consists of a non-empty finite set of objects, called vertices (nodes or points), denoted by V, and a set of edges (arcs, links, branches or lines), each connecting two vertices to each other, denoted by E. The *null graph* is the graph whose vertex set and edge set are empty. Let $v_1, v_2 \in V$. If v_1 is connected to v_2, then $v_1 v_2$ is an edge and v_1 and v_2 are called *adjacent* vertices. Similarly, two edges of graph G are said to be adjacent if they share a common vertex. Edge $v_1 v_2$ is said to be *incident* at vertices v_1 and v_2. A *loop* is an edge connecting a vertex to itself. The degree of a vertex is the number of edges incident to that vertex, where a loop is counted twice. The degree of vertex v is denoted by $\deg(v)$ or $\deg v$.

A *multigraph* is a graph that is permitted to have multiple (parallel) edges (i.e. edges with the same end vertices). A *simple graph* or a *strict graph* contains no loops or multiple edges. In a *directed graph* or a *digraph*, a *direction* might be assigned to each edge. In this case, an edge can be denoted by the ordered pair (v_1, v_2) and therefore (v_1, v_2) and (v_2, v_1) are two distinct edges. The graph is said to be *undirected* otherwise. A graph is *weighted* if a numerical value, called weight, is assigned to each edge of it. This value may be interpreted as cost, length, transfer time, etc., depending on the corresponding context.

Example 2.1   Examples for a simple graph, multigraph, digraph and weighted graph with four vertices are depicted in Figure 2.1. A complete graph and a subgraph are also shown in Figure 2.1.

The number of vertices, $|V|$, is called the *order* of graph G and the number of its edges, $|E|$, is called the *size* of the graph.

Lemma 2.2   *Let $G = (V, E)$ be a simple graph. Then*

$$\sum_{v \in V} \deg(v) = 2|E|. \tag{2.1}$$

The simple graph $G = (V, E)$ is said to be *complete* if for each pair $v_1, v_2 \in V$, the edge $v_1 v_2$ belongs to E. Therefore the size of a complete graph with order m is equal to $n = m(m+1)/2$ and $\deg(v) = m - 1$ for each $v \in V$.

The *complement* of a graph $G = (V, E)$ is defined as $G' = \left(V', E'\right)$ with $V = V'$ such that for each $v_1, v_2 \in V$, $v_1 v_2$ belongs to E' if and only if it does not belong to E. A graph $G' = \left(V', E'\right)$ is said to be a *subgraph* of $G = (V, E)$ if V' is a subset of V and E' is a subset of E. Where $V = V'$, G' is called a *spanning subgraph* of graph G.

The graph $G = (V, E)$ might be represented with an *adjacency matrix*, sometimes or a *connection matrix* with rows and columns labelled by graph vertices, with a 1 or 0 in position (v_i, v_j) if v_i and v_j are adjacent or not, respectively. Specifically, we denote the adjacency matrix of graph G of order m by an $m \times m$ matrix $A = \left[a_{ij}\right]$ where

$$a_{ij} = \begin{cases} 1 & \text{if } (v_i, v_j) \in E, \\ 0 & \text{otherwise.} \end{cases}$$

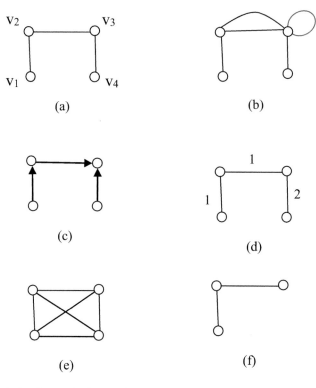

Figure 2.1 (a) A simple graph; (b) a multigraph with one loop; (c) a directed graph; (d) a weighted graph; (e) a complete graph; (f) a subgraph of graph (a).

For a simple graph with no loops, the adjacency matrix must have 0s on the diagonal. For an undirected graph, the adjacency matrix is symmetric. Note that, for multigraphs, a_{ij} is equal to the number of edges between vertices v_i and v_j.

The incidence matrix of graph $G = (V, E)$ with m vertices and n edges is an $m \times n$ matrix $M = [m_{ij}]$ where

$$a_{ij} = \begin{cases} 1 & \text{if } vertex v_i \text{ and edge } e_j \text{ are incident,} \\ 0 & \text{otherwise.} \end{cases} \tag{2.2}$$

Example 2.3 The adjacency matrix of simple graph (a) in Figure 2.1 can be written as follows:

$$A = \begin{bmatrix} 0 & 1 & 0 & 0 \\ 1 & 0 & 1 & 0 \\ 0 & 1 & 0 & 1 \\ 0 & 0 & 1 & 0 \end{bmatrix}. \tag{2.3}$$

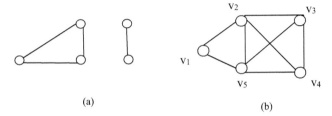

(a)   (b)

Figure 2.2   (a) A disconnected graph with two components; (b) a connected graph.

Also, let $e_i = v_i v_{i+1}, i = 1, \ldots, 3$. The incidence matrix of graph (a) is:

$$M = \begin{bmatrix} 1 & 0 & 0 \\ 1 & 1 & 0 \\ 0 & 1 & 1 \\ 0 & 0 & 1 \end{bmatrix}. \tag{2.4}$$

It can be seen that the degree of each vertex is the sum of its corresponding row entries in either matrix.

A *walk* is an alternating list $v_0, e_1, v_1, e_2, v_2, \ldots, e_k, v_k$ of vertices and edges such that for $1 \leq i \leq k$ the edge e_i has endpoints v_{i-1} and v_i. A u, v-walk is a walk with first vertex u and last vertex v. A *trail* is a walk with no repeated edge. A u, v-trail is a trail with first vertex u and last vertex v. A *path* is a trail with no repeated vertex. A u, v-path is a path with first vertex u and last vertex v. If a walk or trail have the same endpoints $u = v$, then we say they are closed. A closed trail is called a *circuit* and a circuit with no repeated vertices is called a *cycle*. We can denote a path by $v_0, v_1, v_2, \ldots, v_k$ since there are no repeated vertices in the sequence.

The graph $G = (V, E)$ is *connected* if there exists a path between each two vertices of it. Otherwise G is disconnected. If G is not connected then it can be decomposed into several connected subgraphs, which are called *components* of it. More precisely, a *component* is a maximal connected subgraph of a graph.

Example 2.4   In Figure 2.2a a disconnected graph with two components is depicted. Figure 2.2b is connected since each two vertices locate on a path of the graph. For instance, v_1, v_2, v_3, v_5, v_4 is a path in this graph and v_1, v_2, v_5, v_1 is a cycle. A walk is $v_1, v_1 v_2, v_2, v_2 v_4, v_4, v_4 v_3, v_3, v_3 v_5, v_5, v_5 v_2, v_2$. This walk has no repeated edge and so it is a trail.

An isomorphism from a simple graph $G_1 = (V_1, E_1)$ to the simple graph $G_2 = (V_2, E_2)$ is a 1–1 map f from set V_1 onto set V_2 such that $v_1 v_2 \in E_1$ if and only if $f(v_1) f(v_2) \in E_2$. We say G_1 is isomorphic to G_2, written $G_1 \cong G_2$, if there is an isomorphism from G_1 to G_2. It is clear that if $|V_1| \neq |V_2|$ or $|E_1| \neq |E_2|$, then G_1 and G_2 cannot be isomorphic. In other words, only the graphs of the same order and same size can be isomorphic.

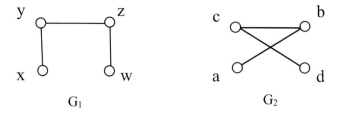

Figure 2.3   Two isomorphic graphs G_1 and G_2

In the case that two graphs are of the same order and size, it is not easy at first sight to recognize if they are isomorphic. One way to find an isomorphism between two graphs is to write their adjacency matrices. If we can write the adjacency matrices with vertices ordered in such a way that the matrices are identical, the corresponding graphs are isomorphic.

Example 2.5   An example of isomorphic graphs is shown in Figure 2.3, where x, y, z and w are mapped onto a, b, c and d. An adjacency matrix for graph G_1 is as follows:

$$A_1 = \begin{array}{c} \\ x \\ y \\ z \\ w \end{array} \begin{array}{cccc} x & y & z & w \\ \left[\begin{array}{cccc} 0 & 1 & 0 & 0 \\ 1 & 0 & 1 & 0 \\ 0 & 1 & 0 & 1 \\ 0 & 0 & 1 & 0 \end{array}\right] \end{array} \tag{2.5}$$

Two adjacency matrices for graph G_2 are A_{21} and A_{22}:

$$A_{21} = \begin{array}{c} \\ a \\ c \\ b \\ d \end{array} \begin{array}{cccc} a & c & b & d \\ \left[\begin{array}{cccc} 0 & 0 & 1 & 0 \\ 1 & 0 & 1 & 1 \\ 1 & 1 & 0 & 0 \\ 0 & 1 & 0 & 0 \end{array}\right] \end{array}$$

$$\tag{2.6}$$

$$A_{22} = \begin{array}{c} \\ a \\ b \\ c \\ d \end{array} \begin{array}{cccc} a & b & c & d \\ \left[\begin{array}{cccc} 0 & 1 & 0 & 0 \\ 1 & 0 & 1 & 0 \\ 0 & 1 & 0 & 1 \\ 0 & 0 & 1 & 0 \end{array}\right] \end{array}$$

As can be seen above, we have: $A_1 = A_{22}$ and so $G_1 \cong G_2$.

A simple graph is called a *bipartite graph* or a *bigraph* (Figure 2.4) if the set of graph vertices can be decomposed into two disjoint sets such that no two vertices within the same set are adjacent. It can be proved that a graph is bipartite if and only if it has no odd cycle. A cycle is called odd if the number of its edges is odd and it is even otherwise. For a simple proof of Theorem 2.6, see [1, page 421].

Theorem 2.6   Graph $G = (V, E)$ is bipartite if and only if it does not contain any cycle of odd length.

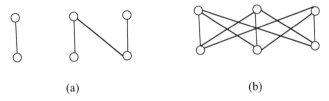

(a)   (b)

Figure 2.4   (a) A bipartite graph; (b) a complete bipartite graph.

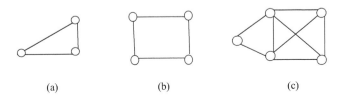

(a)   (b)   (c)

Figure 2.5   Colourable graphs of Example 2.8.

Example 2.7   Consider r jobs and s people where not all people are qualified for all jobs. Can we fill the jobs with qualified persons? This situation can be modelled using a bipartite graph.

The terminology of using *colours* for vertex labels goes back to the map colouring. Here, we consider the issue of vertex colouring. In this regard, colouring using at most k colours is called a (proper) k-colouring. The smallest number of colours needed to colour a graph G is called its *chromatic number*, and is often denoted by $\gamma(G)$. A graph with k-colouring is called k-colourable. Moreover, it is said to be k-chromatic if its chromatic number is exactly k. A subset of vertices assigned to the same colour is called a *colour class* and every such class forms an independent set. Thus, a k-colouring is the same as a partition of the vertex set into k independent sets, and the terms k-partite and k-colourable have the same meaning.

As said before, an *edge colouring* of a graph is a proper colouring of the *edges*, in such a way that no vertex is incident to two edges of the same colour. An edge colouring with k colours is called a k-edge-colouring. The smallest number of colours needed for an edge colouring of a graph G is the *chromatic index*, or *edge chromatic number*. A more general type of colouring is *total colouring*, which includes colouring of both vertices and edges of the underlying graph. Generally, in a total colouring all the vertices and edges are assigned with some colours in such a way that no adjacent vertices, no adjacent edges, and no edge and its end-vertices are assigned the same colour. The total chromatic number of a graph G is the fewest colours needed in any total colouring of G.

A graph G is said to be *planar* if it can be embedded in the plane, or in other words, if it can be drawn on the plane in such a way that its edges do not cross each other, except for their end points.

Example 2.8   Let $G = (V, E)$ be an arbitrary graph with $|V| = m$, then G is evidently m-colourable. Now consider three graphs, G_1, G_2 and G_3, as in Figure 2.5.

It can be seen that G_1 (Figure 2.5a) is not 2-colourable, but for each k, where $k \geq 3, G_1$ is k-colourable and its chromatic number is equal to 3. Moreover, its edge chromatic number is also equal to 3. Also G_2 (Figure 2.5b) is 2-colourable, and its chromatic number is equal to 2.

The edge chromatic number of graph G_2 is equal to 2. Finally, G_3 (Figure 2.5c) is 4-colourable with the chromatic number equal to 4 and the edge chromatic number equal to 6. It can be seen that graphs G_1 and G_2 are planar, whereas G_3 is not.

2.3   Eulerian and Hamiltonian Graphs

The graph $G = (V, E)$ is *Eulerian* if it has a circuit containing all the edges. An *even graph* is a graph in which each vertex has an even degree. A vertex is *odd* if its degree is odd. A graph that is not *Eulerian* is said to be *non-Eulerian*.

Theorem 2.9   Graph $G = (V, E)$ is Eulerian if and only if it is connected and even.

Proof   See [2, page 51].   □

Example 2.10   The graph shown in Figure 2.6 is not Eulerian according to Theorem 2.6. This graph models the Seven Bridges of Königsberg problem. The four vertices show the four districts of the city. The seven bridges are the edges between these four vertices.

As was mentioned before, the question was whether it is possible walk though all four parts of the city while crossing each of the seven bridges only once and returning to the start point. Is there a circuit passing through all edges of this graph? By Theorem 2.6 the answer to this question is negative and therefore the graph is not Eulerian.

The graph $G = (V, E)$ is a Hamiltonian graph if it possess a Hamiltonian cycle (Figure 2.7). A Hamiltonian cycle is a cycle that passes through all vertices of the graph. A graph that is not Hamiltonian is said to be non-Hamiltonian.

A *graphic sequence* $d_1, d_2, \ldots, d_k$ is a non-increasing sequence of positive integer numbers with the property that it corresponds to the degrees of vertices of a graph. Theorem 2.11 provides a simple test for examining whether a sequence is graphical.

Theorem 2.11 (Havel and Hakimi)   A degree sequence $d_1, d_2, \ldots, d_k$ with $k \geq 3$ and $d_1 \geq 1$ is graphical if and only if the sequence $d_2 - 1, d_3 - 1, \ldots, d_{d_1+1} - 1, d_{d_1+2}, \ldots, d_k$ is graphical.

Example 2.12   The sequence $3, 3, 3, 2, 2$ is not graphical because, based on Lemma 2.2, the number of odd vertices in a graph should be an even number, but there are three odd numbers in this sequence. The sequence $3, 3, 3, 2, 1$ is graphical since the sequence $2, 2, 1, 1$ is graphical, because it is the sequence of vertices degrees of graph G_1 in Figure 2.3.

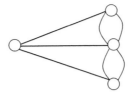

Figure 2.6   The graph of the Seven Bridges of Königsberg problem.

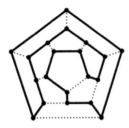

Figure 2.7   A Hamiltonian graph with a Hamiltonian cycle.

2.4   Trees

A graph $G = (V, E)$ is said to be a *tree* if it is undirected and for each two distinct vertices of G there exists exactly one path. In other words, a tree is an acyclic connected graph. For example, Figure 2.1a is a tree. Figure 2.4a is not a tree because it is not connected, and Figure 2.1b is not a tree because it is not acyclic.

A disjoint union of trees is called a *forest*. A *leaf* is a vertex with a degree of 1. From the definition, it is immediately implied that a tree has no loop. Another basic property of a tree is given in Theorem 2.14. We first present Lemma 2.13.

Lemma 2.13   *Let $G = (V, E)$ be a tree with $|V| = m$ and $|E| = n$. Then $n = m - 1$.*

Based on Lemma 2.13, Theorem 2.14 can be easily established.

Theorem 2.14   Let $G = (V, E)$ be a graph with $n \geq 1$ vertices. The following statements are equivalent.

    a.   G is connected and has no cycles.
    b.   G is connected and has $n - 1$ edges.
    c.   G has $n - 1$ edges and no cycles.
    d.   For $uv \in V$, G has exactly one u, v-path.

Theorem 2.14 states that each choice of two properties of the first three defines a tree. Also, we can conclude that cutting one edge from a tree makes it disconnected and adding one edge to a tree forms exactly one cycle. Also, every connected graph contains a spanning tree. A *bridge* in a graph is an edge, where cutting it from the graph make it disconnected. Therefore, every edge in a tree is a bridge.

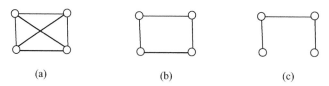

(a)   (b)   (c)

Figure 2.8   (a) A graph; (b) a spanning graph; (c) a spanning tree.

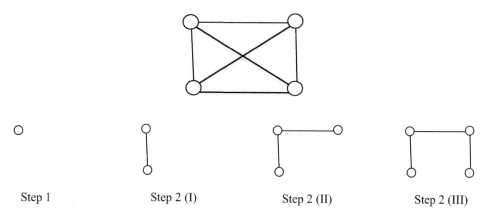

Step 1   Step 2 (I)   Step 2 (II)   Step 2 (III)

Figure 2.9   The output of the vertex-centric algorithm for finding a spanning tree.

2.5   Spanning Trees: Definition and Algorithms

A *spanning subgraph* of graph $G = (V, E)$ is a subgraph with vertex set V. A *spanning tree* (Figure 2.8c) is a spanning subgraph that is a tree. A spanning graph does not need to be connected.

A connected graph can have more than one spanning tree and all spanning trees of it have the same number of vertices and edges. A disconnected graph has no spanning tree. A tree has only one spanning tree that is itself. A tree is a connected forest and every component of a forest is a tree. A graph with no cycle has no odd cycle and so by Theorem 2.6, trees and forests are bipartite graphs. There are many algorithms to compute a spanning tree for a connected graph.

A vertex-centric algorithm for finding a spanning tree for the connected graph G:

1. Pick an arbitrary vertex and mark it as being in the tree.
2. Repeat until all vertices are marked as in the tree:
   Pick an arbitrary vertex u in the tree, which is connected to vertex v not in the tree with edge uv. Add uv to the spanning tree and mark v as in the tree.

We repeat Step 2 for $n - 1$ times since there are $n - 1$ vertices to be added to the tree to obtain a spanning tree.

Example 2.15   Consider graph G in Figure 2.9. The second step of this algorithm has three iterations as shown in Figure 2.9, resulting in one spanning tree shown in Figure 2.9 (Step 2 (III)).

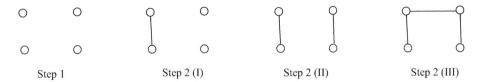

| Step 1 | Step 2 (I) | Step 2 (II) | Step 2 (III) |

Figure 2.10   The output of the edge-centric algorithm for finding a spanning tree.

An edge-centric algorithm for finding a spanning tree for the connected graph G is as follows:

1. Start with a collection of singleton trees, each with exactly one vertex.
2. As long as we have more than one tree, connect two trees together with an edge.

This second algorithm also performs n steps, because it has to add $n-1$ edges to the trees until we have a spanning tree.

Example 2.16   Considering the graph of Example 2.15, we apply the edge-centric algorithm to find the spanning tree shown in Figure 2.10. Step 2 of the algorithm is repeated three times until the resulting spanning tree is obtained.

2.6   Minimum Spanning Tree

Consider the connected weighted graph $G = (V, E)$ with n vertices. All spanning trees of this graph have $n-1$ edges. The *minimum spanning tree (MST)* is the one that has minimum sum of the edge weights. A well-known algorithm to find an MST is Kruskal's algorithm (Figure 2.11). The input of this algorithm is a weighted connected graph and the idea is to maintain an acyclic spanning subgraph H, enlarging it by edges with low weight to form a spanning tree and considering edges in non-decreasing order of weight, breaking ties arbitrarily.

2.6.1   Kruskal's Algorithm for Finding an MST

Consider the connected weighted graph $G = (V, E)$. Find the subgraph H of G through the following steps:

1. Set $V(H) = V$ and $E(H) = \emptyset$.
2. If the next cheapest edge joins two components of H, then include it. Otherwise discard it. Terminate when H is connected.

Theorem 2.17   In a connected weighted graph $G = (V, E)$, Kruskal's algorithm constructs a minimum spanning tree.

Proof   See [3, page 520].   □

Example 2.18   Consider the weighted graph shown in Figure 2.11.

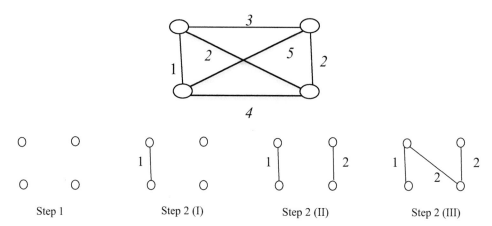

Step 1   Step 2 (I)   Step 2 (II)   Step 2 (III)

Figure 2.11   The output of Kruskal's algorithm for finding a minimum spanning tree.

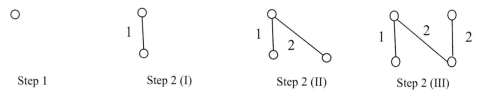

Step 1   Step 2 (I)   Step 2 (II)   Step 2 (III)

Figure 2.12   The output of Prim's algorithm for finding an MST.

Kruskal's algorithm constructs a minimum spanning tree as shown in Figure 2.11.

2.6.2   Prim's Algorithm for Finding an MST

Consider the connected weighted graph $G = (V, E)$. Find the minimum spanning tree of G through the following steps:

1. Choose an arbitrary vertex u and mark it as being in the tree.
2. While the tree does not contain all vertices of G, find an edge with the minimum weight leaving the tree, and add it to the tree.

Example 2.19   Consider the connected weighted graph of Example 2.18. The Prim's algorithm finds a minimum spanning tree of it shown in Figure 2.12.

2.7   Matching in Graphs

Consider a graph $G = (V, E)$. Two edges in G are said to be *independent* if they have no vertex in common. Each subset M of E whose elements are independent edges is called a *matching*. To be more clear, a subgraph $G_M = (V, M)$ of G is a matching if $\deg_{G_M}(v) \leq 1$, for all $v \in V$. In a matching M, if $\deg_{G_M}(v) \leq 1$, then v is said to be *matched* or *covered* by M. Each vertex v that is not covered by M is called *exposed*.

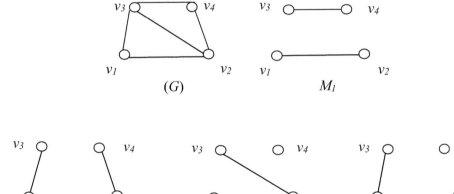

Figure 2.13   Illustration of four matchings in graph G.

Example 2.20   For the graph G shown in Figure 2.13, four matchings are illustrated as M_1, M_2, M_3 and M_4.

From the definition, it can be easily verified that a matching M of $G = (V, E)$ can have at most $\frac{|E|}{2}$ edges. A matching M of graph G is said to be *maximal* if no other edges of G can be added to M. A *maximum matching*, which is also known as the *largest maximal matching*, is defined as the maximal matching which has the maximum number of edges. For a graph G, the *matching number* of G, denoted by $\mu(G)$, is defined as the maximum size of its matchings, that is $\mu(G)$ is equal to the cardinality of a maximum matching of G. A matching M is called a *perfect matching* if it covers all vertices $v \in V$.

Example 2.21   Considering Figure 2.13, it can be observed that each M_1, M_2 and M_3 are maximal matchings by definition. Also, M_1 and M_2 are only maximal matchings, but not maximum. Moreover, M_1 and M_2 are perfect matchings.

A maximum matching of graph G is not necessarily perfect. If graph G has a perfect matching, then the number of its vertices, $|V|$, is even since, if it is odd, and we pair each two distinct vertices, there remains a single vertex that cannot be paired with any other vertex with degree 0. It clearly violates the perfect matching principle.

The concept of matching in bipartite graphs is of great importance. It has many interpretations in different areas of combinatorics, such as the System of Distinct Representatives (SDR). In the following, we consider this issue in more detail.

Let $G = (V, E)$ be a bipartite graph, where V_1 and V_2 define the two parts of V, with the property that $V_1 \cap V_2 = \emptyset$, $V_1 \cup V_2 = V$, and no two vertices in the same part are adjacent. For simplicity, we denote G by $G = ((V_1 V_2), E)$. If $|V_1| \leq |V_2|$, then the cardinality of a maximum matching is at most $|V_1|$. The question that immediately arises here is whether there exists a matching of G that covers all vertices of V_1.

Hall found an answer to this question in 1935. In fact, he provided a necessary and sufficient condition for the existence of such a matching for a bipartite graph. To present Hall's Theorem, we first need to introduce a notation. For each $v \in V$, the *neighbourhood* of v, denoted by $N(v)$, is the set of all vertices in V adjacent to v. Similarly, for each $A \subseteq V$, the neighbourhood of A, denoted by $N(A)$, is defined as $N(A) = \bigcup_{v \in A} N(v)$. Based on this notation, Hall's Theorem is presented in what follows. The reader can find a proof of this theorem in [2, page 150].

Theorem 2.22 Let $G = ((V_1 V_2), E)$ be a bipartite graph, with $|V_1| \leq |V_2|$. Then G has a matching which covers V_1 if and only if $|N(S)| \geq |S|$ for all $S \subseteq V_1$.

It is worthwhile to mention that Hall's Theorem can be generalized for a matching of any size, as follows. A simple proof can be found in [1].

Theorem 2.23 Let $G = ((V_1 V_2), E)$ be a bipartite graph and $k \in N$. Then G has a matching of size k if and only if $|N(S)| \geq |S| - |V_1| + k$, for all $S \subseteq V_1$.

The next corollary is immediately concluded.

Corollary 2.24 Let $G = ((V_1 V_2), E)$ be a k-regular bipartite graph ($k \geq 1$). Then G has a perfect matching.

Another important issue related to matching is the concept of *cover* of edges. Assume that $G = (V, E)$ is a given graph. A subset $W \subseteq V$ is said to be a cover of E (or G) if for each edge $(uv) \in E$, we have $u \in W$ or $v \in W$. For a complete graph K_n, a set W of its vertices is a cover of K_n if and only if $|W| = n - 1$, since for each two distinct vertices of K_n, there exists an edge connecting them. Similarly, for a bipartite graph $G = ((V_1 V_2), E)$, each sets V_1 and V_2 can be considered a natural cover of G. For a graph G, the smallest number of vertices in a cover of G is denoted by $c(G)$. Theorem 2.25 provides a relationship between the two concepts of cover and matching in a bipartite graph. A simple proof is provided in [1, page 491].

Theorem 2.25 For a bipartite graph $G = ((V_1, V_2), E)$, we have $c(G) = \mu(G)$, that is the largest number of edges in a matching is equal to the smallest number of vertices in a cover.

Note that Theorem 2.25 does not necessarily hold true for an arbitrary graph. As if we consider a complete graph K_n with n vertices, it can be verified that $c(K_n) = n - 1$, whereas $\mu(K_n) = [\frac{n}{2}]$. Therefore, it is observed that for $n \geq 3$ we have $c(K_n) > \mu(K_n)$.

Example 2.26 For the bipartite graph G depicted in Figure 2.4b, we have $c(G) = \mu(G) = 2$.

2.8   Network Flows, Max-Flow Min-Cut Theorem

A directed graph (digraph) $G = (V, E)$ in which each edge $e \in E$ is associated with a positive capacity $c(e)$ and two nodes s and t are, respectively, distinguished as

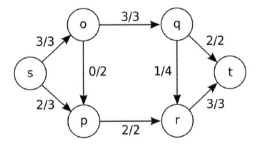

Figure 2.14   A network with a maximum flow.

source and sink is called a *flow network*. In a flow network $G = (V, E, c, s, t)$ each node $v \in Vv \neq st$, is called an internal node.

Network flow problems are basically considered as an important class of problems in operations research, computer science, and engineering, as they arise in many real-life applications. For instance, to model transporting items between locations through routes with limited capacities, a flow network can be modelled. The *maximum flow* and *minimum cost* problems are intrinsically considered as classic problems in the network flow field, which are briefly introduced in the following, and well-known algorithms are provided for them.

2.8.1   Maximum Flow Problem

In a flow network $G = (V, E, c, s, t)$ the aim of the maximum flow problem is to find the total amount of flow from source s to sink t in such a way that the flow on edge e does not exceed $c(e)$ for all $e \in E$ and for every internal node v the incoming flow to the node is equal to the outgoing flow from it. Such a flow is called an admissible flow. Therefore, in a maximum flow problem the task is to find the maximum admissible flow.

Example 2.27   Consider the flow network depicted in Figure 2.14. The source and sink are denoted by s and t respectively. This network has four internal nodes. There is a fraction associated to each arc of this network. The denominator is the capacity of that arc and the numerator shows the flow passing through it, which is evidently at most equal to its capacity. As can be seen in this figure, the maximum flow of 5 is transmitted from source s to sink t.

In Example 2.27, a maximum flow of 5 in the network is illustrated. However, the main question that arises here is how we can verify that no better flow exists in the given network. In order to find a criterion that answers this question, we need to be familiar with the second network flow problem, called the minimum cut problem.

2.8.2   Minimum Cut Problem

In a flow network $G = (V, E, c, s, t)$, a subset of edges C is called a cut if after removing all edges in C from the network G, there does not exist any (directed) path from the source s to the sink t. The cost of removing each edge $e \in C$ is equal to the capacity of

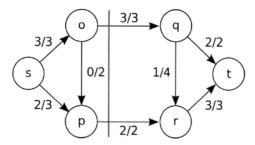

Figure 2.15 A minimum cut for the network of Figure 2.14.

that edge: $c(e)$. Finally, the total cost of a cut C is defined as the sum of the capacities of all its edges in C. Given a flow network $G = (V, E, c, s, t)$, the aim of the minimum cut problem is to find a cut with the minimum total cost.

Theorem 2.28 provides the relationship between the two network flow problems defined here. For a general proof see [3, page 184].

Theorem 2.28 The maximum (admissible) flow value in a network is equal to the total cost of the minimum cut in that network.

Example 2.29 Looking back to the network given in Figure 2.14, one can observe that $C = \{(o, q), (pr)\}$ is a cut with the total cost equal to $3 + 2 = 5$. Figure 2.15 illustrates this cut. Therefore, it is concluded that the maximum flow of the corresponding network should be equal to 5, that is the flow given in Figure 2.14 is exactly the maximum flow.

References

[1] Brualdi RA. *Introductory Combinatorics.* New York: Pearson, 1992.

[2] Bondy JA, Murty USR. *Graph Theory with Applications.* London: Macmillan, 1976.

[3] Ahuja RK, Magnanti TL, Orlin JB. *Network Flows.* New York: Pearson, 1993.

PART II
NETWORK ANALYSIS

3 Structural Analysis of Biological Networks

Narsis A. Kiani and Mikko Kivelä

3.1 Introduction

The theory of complex networks plays an essential role in a wide variety of disciplines, ranging from communications to molecular and population biology.

The behaviour of most complex systems, from the cell to the Internet, emerges from the orchestrated activity of many interacting elements. Each area, naturally, has its unique questions and problems under study. Nevertheless, from a statistical perspective, there is a methodological foundation that has emerged, composed of tasks and tools that are each common to some non-trivial subset of research areas involved with network science. At a highly abstract level, the elements can be reduced to a series of nodes that are connected by links, with each link representing the interactions between two components.

The focus of this chapter is on analysis of the biological network. Network data are collected daily in a host of different areas. Biological interactions at many different levels, from the atomic interactions to the relationships of organisms in an ecosystem, can be modelled as networks. Physical interactions between molecules, such as protein–protein interactions, can easily be conceptualized using the node–link terminology and form a network that represents the system (Figure 3.1). Depending on the nature of the interactions, networks can be directed or undirected. For example, in the direction of modulation of a gene by another gene, or the direction of information flow from a transcription factor to the gene that it regulates, the interaction between any two nodes has a clear direction. In undirected networks, the links do not have an assigned direction. For example, in protein interaction networks, a link represents a shared binding relationship: if protein X binds to protein Y, then protein Y also binds to protein X.

It is possible to categorize many of the various tasks faced in the analysis of network data across different domains according to a statistical taxonomy. It is along the lines of such a taxonomy that this chapter is organized, progressing from descriptive methods to methods of modelling and inference. We will survey methods and approaches in network theory, along with current applications in

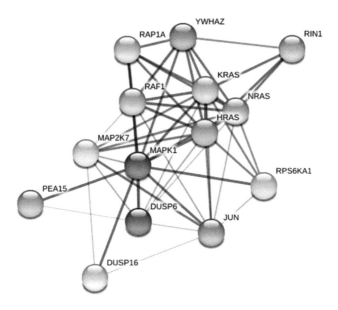

Figure 3.1 An example of a biological network. Here, the graph represents interaction among essential components of the MAP kinase signal transduction pathway. The interaction has been found using the STRING Database.

biomedical informatics. We review different procedures for the measurement of centrality and importance in biological networks. To find the most important nodes in a large complex network is of fundamental importance in computational biology. Then we review how motifs and modules in biological networks can be identified. Our primary goal in this chapter is to provide as broad a survey as possible of the significant advances made in this field, and we hope that this chapter will serve as a useful introduction to the area for those unfamiliar with the literature.

3.2   Structural Properties of Biological Networks

In the study of a complex system, questions regarding some aspect of the structure or characteristics of a corresponding network have proven very useful. For example, finding a protein complex may be addressed as a graph-partitioning problem; questions involving the movement of information or spread of a disease can be modelled in terms of paths and flows along those paths on a network. The primary goal here is to connect the topological features of biological networks with biological function, design principles of regulation mechanisms and evolution of the systems. For example, it has been suggested that prominent structural features, such as the shape of the degree distribution, could be explained by a simple network growth model [1] or that protein complexes as discrete units of the function of a system can be found by community detection methods [2]. The tools commonly used for the structural analysis of cellular networks are the same ones that are used within network science for a multitude of other types of networks, ranging from social network to transportation networks and many others. These methods shed light on the topologies of the cellular networks and their functional organizations. We present in this section a brief overview of some of the many such tools available.

3.2.1 Node and Links Characteristics

Nodes (vertices) and links (edges) are fundamental elements of a network, and there are several network characterizations centred upon these. We discuss several such characterizations in this section. We present the following three aspects of network structure: (1) degree distributions; (2) connectivity; and (3) node centrality measures.

Degree Distributions
The degree dv of a node v in a network $G = (V, E)$ counts the number of links in E incident upon v. Much of the recent research on the structure of biological and other real networks has focused on determining the form of their degree distributions, which measures the proportion of nodes in the network having degree dv.

Formally, given a network G, we define P_k to be the fraction of vertices $v \in V$ with degree $dv = k$. The collection $\{P_k\}k \geq 0$ is called the degree distribution of G. Degree distribution is a rescaling of the set of degree frequencies, computed from the original degree sequence [3, 4]. The degree distribution carries crucial information about the network, its evolution and formation process, and can exhibit a variety of shapes. For example, network growth processes with a preferential attachment mechanism can create degree distributions with power-law form

$$P(k) \sim k^{-\gamma} \tag{3.1}$$

Several real-world networks, including protein–protein interaction networks, have been reported to be scale-free networks, that is, to contain a power-law degree distribution. However, these findings have been questioned in later research with the claim that other fat-tailed distributions, such as log-normal distributions, would be a better fit to the data [5].

The topic of the exact shape of degree distributions remains controversial [6], but it is clear that many real-world networks have broad-tailed degree distributions. This means that these networks have many more high-degree nodes than, for example, in a fully random (Erdös–Rényi) network with the same number of links. That is, they contain *hubs*: nodes that have a significantly higher degree than the average one in a network. These high-degree nodes have received considerable attention and have shown a tendency to play an essential role in protein–protein interaction networks [7, 8]. Luscombe et al., in the yeast transcriptional network, found that 'a few transcription factors serve as permanent hubs but most act transiently only during certain conditions' [9].

The hubs can be further classified if one considers the timings and locations of the connections. Han et al. define two different types of hubs in protein–protein interaction networks: party hubs and date hubs [8]. Party hubs interact with all or most of their neighbours simultaneously, whereas date hubs bind their neighbours at different times or locations. Ekman et al. [10] have shown that party hub proteins contain

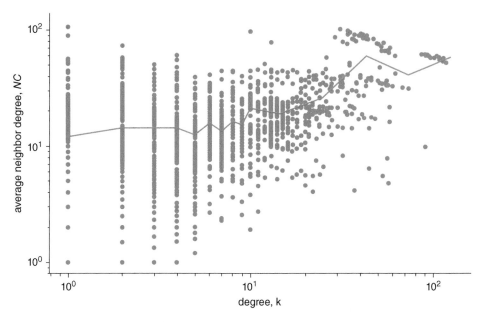

Figure 3.2 Average neighbourhood connectivity versus degree for each node in the yeast protein–protein interaction network. The solid line represents the average nearest neighbour degree for a degree bin.

long, disordered regions more often than do date hubs. They indicate that disordered regions are important for flexible binding, but less so for static interactions.

It can be interesting to understand the manner in which nodes of different degrees interact with each other. This can be assessed by the average degree of the neighbours of a given vertex, neighbourhood connectivity or neighbourhood connectivity distribution.

In undirected networks, the neighbourhood connectivity of a node n is defined as the average connectivity of all neighbours of a node v [11]:

$$\mathbf{NC(v)} = \left(\sum\nolimits_{u\in\mathbf{N(v)}} |\mathbf{N(u)}|\right)/|\mathbf{N(v)}|,\qquad(3.2)$$

where $\mathbf{N(v)})$ is a set of neighbours of node v.

From the neighbourhood connectivity, one can calculate the average neighbourhood connectivity of all nodes with exactly degree k (i.e. k neighbours) for every possible value of $k = 0, 1, \ldots, K$.

Figure 3.2 shows the neighbourhood connectivity versus node degree for the yeast protein–protein interaction network. This plot suggests that in contrast to nodes of lower degree that tend to connect with nodes of both lower and higher degrees, nodes of higher degrees tend to interact only with nodes of higher degree. For directed networks, one can extend to in- and out-connectivity in analogy to the in- and out-degree in directed networks. The interpretation of neighbourhood connectivity of a node is often domain-specific. For example, in genetic interaction networks, nodes with higher out-degree are more general suppressors than nodes with a lower out-degree. Therefore, the 'only in' neighbourhood of node v shows the universality of the suppressors of v.

Maslov et al. have argued that shared neighbours in protein–protein interaction and regulatory networks increases the overall robustness of the network [11].

It is important to note that many of the results about degree distribution of biological networks, including those cited above, have been carried out on sampled subnetworks rather than on complete networks, and the effect of sampling on these findings has been questioned by some researchers [12, 13].

Connectivity

Connectivity is one of the basic concepts of graph theory and network science. As you recall from Section 3.1, a path is a sequence of edges that begins at a node of a network and travels along the edges of the network, always connecting pairs of adjacent nodes. If there is a path between a pair of nodes, they are said to be connected. An undirected network can be divided into connected components in which all the pairs of nodes are connected (and between the connected components none of the nodes is connected). A network with only a single connected component is said to be connected.

Depending on the context, *path length* may either be the number of edges on the path or the sum of the weights of the edges on the path. For example, in the graph shown in Figure 3.3a, we have a path from 1 to 8 with a length of 4. The same path but on the weighted graph has length 20, as shown in Figure 3.3b. Since a network may have more than one path between two nodes, we may be concerned about finding a path with a particular property. For example, finding a path with minimum length defines a *distance* between the two nodes (sometimes called the geodesic distance). It is also possible that we cannot always find a path between every two nodes of the graph. The average shortest path length gives the expected distance between two connected nodes. Additional insight into the shortest path lengths is provided by shortest path length distribution. The greatest distance between two nodes of a network is called network diameter. If a network is disconnected, its diameter is the maximum of all diameters of its connected components. The basic idea of connectivity is reachability among nodes of a network by traversing its edges. An undirected graph is connected if there is a path between every pair of its vertices. The minimum number of edges that need to be removed to disconnect the remaining nodes from each other in the graph is an important measure in the study of the resilience capacity of a network. The average path lengths and diameters of biological networks are small in comparison to network size [14, 15]. For example, the average path lengths of metabolic networks were found to be, on average, approximately three edges. Similar findings have been reported for protein and genetic interaction networks, where the average path length ranges from about 4 to 8 [16, 17].

The average path length in a network can be seen as an indicator of how freely information can be conveyed through it. Thus, we can say the findings of the short average path in biological networks suggests that such networks are efficient in the transferring of biological information. Only a small number of intermediate interactions are necessary for any one protein/gene/metabolite to influence the behaviour of another. Note that random networks also have short distances and that in real

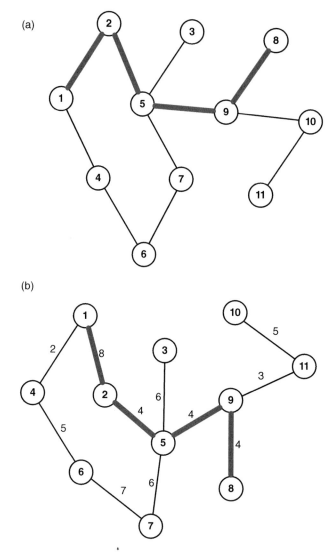

Figure 3.3 The path from 1 to 8 on two graphs is demonstrated with a red line. The length of the path is 4 in (a), while in the weighted graph (b), the length of the same path is 20.

networks, short distances are also more of a rule than an exception. This is because, in any network, a very small number of uniformly randomly placed connections is enough to create shortcuts that collapse the path lengths to small numbers [14, 15]. This implies that any network with long path lengths should be created under severe (spatial or other) constraints forbidding the appearance of most links.

Node Centrality
We often have reason to believe that the element at the centre of every system is essential. For instance, centrality measures have been used to predict the essentiality of a gene or protein based on network topology [18]. To date, a vast number of different centrality measures have been devised for ranking the nodes in a complex network

and quantifying their relative importance. We have already learned about what is arguably the most widely used measure of node centrality: node degree.

Nodes with a large number of neighbours (i.e. edges) have high degree centrality. Degree centrality, however, is only indicative of the centrality of the node in the wider network, because it is a purely local measure.

The global centrality of a node is an ambiguous concept, and it can be seen from several aspects. Here we will focus our discussion primarily around three commonly used global node centrality measures that capture a wide range of importance in a network: closeness, betweenness, and eigenvector.

Closeness centrality measures are usually expressed in terms of the distance, $\delta(u, v)$ between nodes in a graph or network. The core idea behind closeness centrality is the following. A node is considered vital if it is relatively close to all other nodes. Nodes with short paths to all other nodes in the network have high closeness centrality.

The standard approach to calculate the closeness centrality of a node [19] is to calculate the inverse of the summation of the distance of each node to every other node in the network:

$$C_c(v) = \frac{1}{\sum\limits_{u \in V} \text{dist}(u, v)},$$
(3.3)

where $\text{dist}(v, u)$ is the distance between the vertices $u, v \in V$. This measure can be normalized with the number of nodes in a network for a more fair comparison across graphs and with other centrality measures.

Betweenness centrality measures are aimed at summarizing the extent to which a node is located 'between' other pairs of nodes. A node, which lies on communication paths, can control communication flow, and is thus important. Betweenness centrality quantifies where a node is located with respect to the paths in the network. The most commonly used betweenness centrality, introduced by Freeman [20, 21] as a way of quantifying an individual's influence within a social network, is defined as:

$$C_b(v) = \sum\limits_{\substack{s \neq t \\ s \neq v \\ v \neq t}} \frac{\sigma_{st}(v)}{\sigma_{st}}$$
(3.4)

where $\sigma_{st}(v)$ is the total number of shortest paths between s and t that pass through v, and $\sigma(s, t)$ is the total number of shortest paths between s and t (regardless of whether or not they pass through v). The mean value of betweenness centrality for the essential proteins in the network has been reported to be approximately 80% higher than for non-essential proteins in the yeast protein interaction network and the mean [22].

Eigenvector centrality measures can be formulated in terms of eigenvector solutions of appropriately defined linear systems of equations. The basic idea behind these measures is that an important node is connected to important neighbours. That is, the centrality $C_{ev}(v)$ is defined recursively based on the centralities of all the neighbouring nodes, such that

$$C_{ev}(v) = \frac{1}{\lambda} \sum\limits_{u \in V} C_{ev}(u)$$
(3.5)

This problem can be rewritten as an eigenvalue problem such that the normalization constant λ corresponds to an eigenvalue and the centralities correspond to the elements of an eigenvector. Eigenvalues are a set of scalars associated with a matrix. Each eigenvalue is paired with a corresponding eigenvector. To obtain the eigenvalues, we should solve the equation $A\vartheta = \lambda\vartheta$, where λ is an eigenvalue and ϑ is an eigenvector. If we solve this equation for the adjacency matrix of a graph, A, then we have an eigenvalue assigned to that graph. In general there can be multiple solutions to this equation and thus multiple eigenvalue and eigenvector pairs. (The set of all eigenvalues is called graph spectrum or graph spectra.) Of specific interest is the solution leading to the largest eigenvalue, which is called the spectral radius of a network:

$$\rho(A) = \text{Max}|\lambda| \tag{3.6}$$

There are many variations of eigenvector centrality measures [23, 24]. For example, Bonacich defines an eigenvector centrality measure as follows. If A is the adjacency matrix of network G, then the eigenvector centrality score, $C_{ev}(v)$ of the node v is given by the corresponding coordinate of a normalized eigenvector ϑ satisfying $A\vartheta = \rho(A)\vartheta$.

For a unique ranking of the nodes in a network, it is necessary that the eigenvalue $\rho(A)$ has a geometric multiplicity of 1. When the network is undirected and connected, the largest eigenvalue of A will be simple and its eigenvector will have entries that are all nonzero and share the same sign [25, 26].

3.2.2 Motifs and Modules in Biological Networks

The measures discussed in the previous section were concerned with identifying individually important nodes within a network. However, several recent studies have revealed that biological networks have a comparatively high number of links in their immediate neighbourhood, that is a high network clustering coefficient, independent of the network size [1, 15]. This property in biological networks might be linked to the existence of a local structure such as groups of loosely connected dense clusters (modules) of individual nodes collaborating to carry out some specific biological function. For example, protein complexes strongly correlate with dense subgraphs in protein–protein interaction networks, and the feed-forward loop corresponds to a motif in biological networks [27, 28]. Associating the topological structure with biological knowledge provides a promising tool to understand the biological mechanisms of species, and this feature has been applied to predict the function of an unknown protein based on the cluster it belongs to [17, 29]. In this section, we will discuss motifs as the building blocks of the biological network. We then consider the problem of community detection in biological networks and describe a number of algorithms that have been developed for this purpose.

Motifs
Milo et al. [28] showed that networks from diverse fields – biological and non-biological – contain several small topological patterns that are so frequent that it is

unlikely to occur by chance. Moreover, different networks tend to have different sets of such persistent local structure. These patterns, referred to as 'network motifs', are recognized as 'the simple building blocks of complex networks'. Motifs are small topological patterns. Formally, a network motif is a small, over-represented partial subgraph of a real network. Here, over-represented means that it is over-represented when compared to networks coming from a random graph model.

A procedure for finding n-node networks motifs goes as follows. First, find all n-node connected subgraphs in the real graph. For example, there are 13 three-node subgraphs, 199 four-node subgraphs and so on. Then find all n-node subgraphs in a set of randomized graphs, which keeps some of the features of the real graph. We discuss the random network models in the next section, and here we deliberately leave out the details of how the randomization is done. However, it is important to realize that the expected frequency of a given subgraph can be largely determined by more simple features of the graph, such as the degree distribution. Therefore, in order to deem a specific subgraph as overrepresented in a particular network due to a specific biological mechanism and not just an artefact of the degree distribution, we must compare its abundance to that of a randomized network with the same degree distribution. Finally, one assigns each of the n-node subgraphs a probability (p-value) for them to occur more at random than in the real graph. One three-node motif and one four-node motif, so-called feed-forward motif and bi-fan motif, have been identified in transcriptional regulatory networks of *E. coli*, *S. cerevisiae* and the neuronal network of the nematode *C. elegans* [28, 30]. In a motif, the directed links can symbolize the activation or the inhibition of the target gene, or any combination thereof. Thus, each undirected motif can have multiple different directed versions. For example, eight different versions of the feed-forward motif can be constructed, each with a different biological function [31]. Although motifs have been shown to be closely related to conventional building blocks [9, 32, 33], it is not clear what dictates the particular frequencies of motifs in a specific network or how the knowledge about distribution of motifs can be used further, for example in network inference or prediction of missing links in biology.

Community (Modules) Detection

Detecting communities, modules or clusters in complex networks is an important problem in many scientific fields, from technology to sociology. This is also true in biological networks. For example, Ravasz et al. [34, 35] have reported that metabolic networks of different organisms are organized into 'many highly connected components that combine in a hierarchical manner into larger less cohesive units'. Detecting communities allows revealing the existence of such non-trivial mesoscale structures of the network and to inspection of the networks at a coarse-grained level. This allows further to infer unique relationships between the nodes that may help to understand better the properties of dynamic processes taking place in a network. For example, in social networks, spreading processes of epidemics and innovation are significantly affected by the community structure of the graph [36–38]. Further, much of the interest towards communities is because they can reflect the other features of the nodes in real data: Communities in a network are sometimes a group of similar nodes and have more common features to each other but are different from the rest of the nodes in the network. However, such correspondence of node metadata and network structure is not always perfect or can even be misleading [39].

There is no universally accepted definition for a community. Qualitatively, a community can be defined as a subset of nodes within the network which are highly connected to each other and sparsely connected to the rest of the nodes in the network [40]. However, a precise definition of a community structure that can be used to develop algorithms for finding such structures can be done in several different ways. These definitions can, and often will, lead to very different structures being found in real networks.

Ideally, the choice of the definition should be motivated by the real-world problem one is trying to solve with community detection, such that the definition matches the goal as closely as possible. Some choices are often obvious based on the problem, such as choosing if the communities can overlap or if all nodes need to be included in a community. However, many of the choices are not as straightforward as this, as the community detection methods are often defined as relatively complicated algorithms or as functions to be optimized. Further, the methods can contain implicit features that might not even be apparent to the creators of the method. An excellent example of such a feature is the *resolution limit* in a prominent community detection method (modularity optimization), which was found after the method already became popular, and meant that it could not find communities smaller than a size dependent on the network size [41].

Various computational methods for community detection have been proposed [42–49]. These methods either propose a fully new definition of a community structure or improve the algorithms for finding an optimal community structure with a given definition. It is important to make this distinction. As discussed above, the definition of a good community structure defines the problem that one wants to solve and is dependent on the goals of the research. The particular algorithm then defines the computational speed (and in many cases the accuracy) at which this problem can be solved. However, to complicate this issue further, sometimes a worse-performing algorithm for the same problem has been seen to produce more desirable results [38, 50].

Community detection methods can be characterized and classified in many ways. As already mentioned, they can either find partitions (divisions of sets to non-overlapping parts) or covers (arbitrary sets of subsets) of all the nodes. Further, one can define communities from a global perspective as an optimization problem in which each partition or cover is given a score of how well it divides a network into communities. In contrast to this, a local perspective would be constructing communities just by inspecting the nodes in a community and their immediate neighbourhoods independent of the other parts of the network. For example, Raghavan et al. [51] proposed an algorithm based on the idea that each node in the network should be in the same community as the majority of its neighbours. This algorithm starts with allocating a distinct community to each node in the network. After that, the nodes in the network are listed in random sequential order through the sequence. Finally, each node takes the label of the majority of its neighbours. Further, community detection methods can then be subdivided to (a) top-down or divisive hierarchical techniques in which a series of network partitions hierarchically decompose a network into modules; or (b) bottom-up or agglomerative hierarchical procedures in which modules are built by adding elements to an initial seed.

An important consideration in the algorithms for community detection is their computational complexity. Algorithmic developments of community detection methods aim to improve the identification of meaningful communities, while keeping the computational complexity of the underlying algorithm as low as possible using optimization of objective functions. Many of the community detection methods are defined in a way that their optimal solution would require one to solve a problem in which the best known algorithms require exponentially growing time as a function of the network size. To remedy this, often heuristic optimization methods are employed to solve these problems at least suboptimally. These type of algorithms are often randomized and lead to a different solution at every run. There are also different compromises of sacrificing some of the optimality to make the algorithm run faster, and different methods might work better depending on the target network.

From a practical point of view, it is often important to try out different methods [52]. If one is using methods that produce different results on each run, it is also essential to run these methods multiple times to check that the end results are robust. Ideally, sanity checks, such as inspecting the network division visually for small networks, should be used. In the case that the network contains a solid community structure, the choice of method might not make a big difference, and all of them should produce similar results. It should also be noted that many methods will report a community structure even in the case that the network does not have any clear structure (e.g. it is a random Erdös–Rényi graph).

It is impossible to catalogue all the community detection methods here or even give in any way a comprehensive or fair review of them. For a recent survey for practical community detection, we refer the reader to the article by Fortunato and Hric [49], and for more systematic reviews the reader should follow the references in that article. However, in the interest of giving the reader more concrete understanding of community detection methods, we will next give a few examples of popular methods that are, important in the history of the field or otherwise interesting.

The Newman and Girvan (NG) algorithm is one of the early algorithms of community detection in graphs [53]. It is a divisive hierarchical algorithm in which edges are iteratively removed based on the value of their betweenness centrality, such that the edges with high centrality scores are removed first. The connected components of the network in which the edges are removed are then interpreted as communities. The procedure can be continued until all the nodes are in their own communities (connected components), and the end result in this case is a hierarchical clustering. In order to choose an optimal level of the hierarchy and thus a single partition of nodes to communities, the authors of this method developed a measure called *modularity*. Modularity measures the number of edges within the communities with the expected number of such edges deduced. The expected number of edges is typically calculated based on the assumption that the degrees of the nodes are kept, but the network is otherwise maximally random (but other reference models are possible). Interestingly, optimizing modularity of a network partition directly has become one of the most popular methods of community detection, much surpassing the popularity of the original NG algorithm. The computational cost of the NG algorithm can be significant, $O(N^3)$ on a sparse graph, stemming from repeated edge betweenness computations [50]. Clauset et al. propose a greedy optimization version of the NG algorithm, which uses more efficient data structures and has a complexity of $O(N \log 2N)$ on sparse graphs [54].

Finding the partition that achieves the highest value of modularity is a challenging computational problem (in fact, it is NP-complete [42–47]). A heuristic algorithm that in some benchmarks has produced good results [46] is the Louvain method [55]. It is a multistep technique based on local optimization of modularity in the neighbourhood of each node. In the first step, this method allocates a different community to each node of the network, and then a node is moved to the community of one of its neighbours in a way that the move maximizes the change in modularity. The above step is repeated until no further improvement can be made. In a second step, a hierarchical community structure is then created by treating each community as a single node and repeating the first step to that network. The second step is repeated until there is only a single node left, or when the modularity cannot be improved in a single step. The computational complexity of the algorithm is linear following the number of edges of the graph [50].

Methods to optimize modularity (or any other global objective function) can, of course, include many other possibilities beyond merely changing the community assignment of a single node – for example, joining two communities or splitting a community into two. Sobolevsky et al. [56] optimize modularity using all of these possibilities. The Louvain algorithm works via greedy optimization where the objective function is always improved, but one can also combine any of these steps with any other optimization strategy, such as simulated annealing [57, 58].

There are other optimization-based algorithms with different objective functions that provide different approaches to solve the community detection problem. Some of these methods are based on more traditional data clustering techniques by transforming the community detection problem to clustering data in a metric space or with some pairwise similarities. For example, some of these methods define the similarity function in high dimensional feature space using spectral clustering. The reason for using eigenvector space is that the projected eigenvectors significantly distinguish the similar nodes into more distanced positions in feature space. The leading eigenvector method is one such approach. The heart of this algorithm is modularity maximization, but the modularity is expressed in terms of the eigenvalues and eigenvectors of a matrix, called the modularity matrix [59]. In this algorithm, first, the leading eigenvector of the modularity matrix is obtained, and then the graph is split into two parts in a way that maximizes modularity, defined based on the leading eigenvector, improvement. Then, at each step in the subdivision of a network, the modularity contribution is calculated. The algorithm stops once the value of the modularity contribution is not positive [38]. The computational complexity of each graph bipartition on a sparse graph is $O(N^2)$ [60].

Although we have here mainly focused on modularity optimization as a prominent example of community detection, there are other approaches. In many applications, it is intuitive to think of communities through processes of flows or diffusion on the network. These flows are trapped within the communities, and this idea can be formalized with random walks. In 2006, Pons and Latapy proposed a new algorithm called Walktrap to detect community structures [61]. It is a hierarchical clustering algorithm that measures the distance of nodes by random walks among nodes. The short distance random walks tend to stay in the same community. The algorithm for sparse networks runs in time $O(N^2 \log N)$ and space $O(N^2)$. Later, Rosvall et al. introduced a novel method that figures out communities in weighted and directed networks by using random walks to analyse the information flow through a network [62, 63]. The network was decomposed into modules by compressing information flow over the network. This algorithm runs in $O(E)$. For an extensive survey of this area and comparison, see [38, 50]

3.2.3   Random Network Models

Informally, a network model is a process (randomized or deterministic) for generating a network. We can think of models of static networks or evolving networks. Models of static networks get a set of parameters Π, and the size of the network N as input; they return a network as an output. Models of evolving networks get a set of parameters Π, and an initial network G_0 and return a network G_t for each time t as output. If the model is deterministic, then it defines a single network for each value of n (or t). In contrast, a randomized model defines a probability space G_n, P where G_n is the set of all networks of size N, and P a probability distribution over the set G_n (similarly for t). We call this a family of random graphs R, or a random graph R.

In 1959, Paul Erdös and Alfred Rényi introduced their now classical notion of a random graph to model non-regular complex networks [64, 65]. In an Erdös-Rényi (ER) random network, each pair of nodes is connected with equal probability. The basic idea of the ER random network model is the following. The $G_{n,p}$ model get the number of nodes n and a parameter $p, 0 \leq p \leq 1$, and for each pair of nodes (i, j), generate the edge (i, j) independently with probability p. In a related but not identical model, one selects m edges uniformly at random to add to the network.

Simple as it may be, the Erdös–Rényi network features some characteristics commonly observed in many real-world biological networks. The degree distribution of an ER network follows a binomial distribution, which for a large and sparse ER network can be approximated by a Poisson distribution. The tails of degree distributions of an ER graph are typically narrow, meaning that the node degrees tend to be tightly clustered around the mean degree. There will be no regions in the network that have a large density of edges. The ability to maintain a flow of information, mass or energy, between all of the nodes in any biological network is crucial, and as you saw in the earlier section, this prerequisite can be translated into the connectivity and the length of network paths [66, 67, 69, 86]. For an ER network, the average path length is strikingly small compared to the network size [58]. This is a beautiful and elegant theory that has been studied exhaustively for the last 60 years; ER graphs had been used as idealized network models, and many deep theorems about the properties of ER graphs have been established. But unfortunately, such fully random networks are not sufficient to model real-world networks [68, 70]. The degree distribution of many real-life networks is fat-tailed [18, 70, 71]. These degree distributions are usually so skewed that it makes sense to plot the histogram in log-log form, where the characteristic distribution then becomes clear: very low degrees are possible, and there are few nodes with a large number of connections. Further, real biological networks have high clustering coefficient and communities, as we discussed earlier. So we need the models that better capture the characteristics of real graphs in terms of degree sequences, the clustering coefficient and other features. Early attempts to model such structures were typically lattices or other regular structures that contained clustering but at the same time suffered from unrealistically long path lengths.

Watts and Strogatz illustrated that the real networks might lie somewhere between randomness and regular structure. They showed that only a small number of random connections is enough to dramatically reduce the paths lengths even in the most strictly structured networks. They built a network model (WS model) by starting from a lattice and randomly rewiring some of its links. This resulted in networks with clustering close to that of a lattice and path lengths similar to those of ER random networks [72]. The approach is based on the idea of the small-world graph, Milgram's famous work [73], where the substantive point is that networks are structured such that even when most of our connections are local, any pair of people can be connected by a small number of relational steps. They measured two elements: the clustering coefficient and the average distance between all nodes in the network. In a highly clustered, ordered network, a single random connection will create a shortcut that lowers the average distance among the nodes in the network dramatically. Watts and Strogatz demonstrated that small-world properties can occur in graphs with a surprisingly small number of shortcuts. Their model matches the key features of large real-world networks, which they found to have very sparse connectivity, large clustering coefficient values (larger than in ER graphs of the same size), and short average path lengths (similar to ER graphs of the same size).

Fat-tailed degree distributions were reported and explained early in citation networks, where it was noticed that some papers receive a very large number of citations while most papers receive very few. In 1965 Price suggested a model for citation networks [74]. The proposed mechanism was as follows: each new paper is generated with m citations, and new papers cite previous papers with probability proportional to the number of citations they already have. This mechanism is motivated by the fact that authors often become aware of articles by finding them in a citation list of another article, and thus the more an article is cited the more likely it will be cited again. More precisely in the Price model, each paper is considered to have a 'default' citation and the probability of citing a paper with degree k is proportional to $k + 1$. The average in-degree distribution of networks generated using this model follow a power-law with exponent $\alpha = 2 + 1/m$. [74, 75] In a later paper [76] in 1976, Price presented a mechanism to elucidate the occurrence of power-laws in citation networks, which he called 'cumulative advantage', but which is today more commonly known under the name preferential attachment.

Many other networks can be considered to have a similar preferential attachment mechanism to citation networks. New links are created via following the old links, and thus the nodes with high degrees gain more links than low-degree ones. For example, in social networks, new friends are found through existing ones and in genetic networks the gene duplication results in copying of links. Decades later, Barabasi and Albert (BA) suggested a model following the idea of preferential attachment [77]. The BA model gets an initial subgraph G_0 and m the number of edges per new node as input. Then nodes arrive one at a time, and each node connects to m other nodes, selecting them with probability proportional to their degree. At the limit of the infinitely large network size, the result is a network with a power-law tail in the degree distribution with exponent $\alpha = 3$.

3.3   Modelling of Biological Networks

3.3.1   Network Inference

Biological interaction networks provide a natural framework for describing different aspects of biological systems, such as a living cell. Various classes of biological networks have been introduced and studied. Transcriptional regulatory networks, protein interaction network and metabolic networks are among the most well-studied biological networks. In biology, transcriptional regulatory networks and metabolic networks would usually be modelled as directed graphs. Metabolic networks combine the knowledge on metabolic pathways and cycles in an organism or a cell type. Nodes in the metabolic networks are metabolic substrates or products, and edges represent transitions through chemical reactions, catalysed by enzymes. Transitions are always directional, and thus the metabolic networks are directed. In expression networks, nodes are genes, and edges connect genes that are coexpressed. These networks are constructed by large-scale DNA microarray experiments, and the unordered composition of a pair of coexpressed genes leads to the undirected nature of the networks. There are two types of nodes in regulatory networks – genes and transcription factors. An edge connecting a gene and a transcription factor indicate up- or downregulation. In a transcriptional regulatory network, nodes would represent genes with edges symbolizing the transcriptional relationships between them. The transcriptional regulatory network would be a directed graph because, if gene A regulates gene B, then the edge naturally starts at A and terminates at B [78]. Genetic interaction networks are another class of biological network. In general, two genes are said to interact if a mutation in one gene either suppresses or enhances the phenotype of a mutation in the other gene [79]. The nodes in genetic interaction networks are genes, and the edges are genetic interactions. Containing only directed interactions, genetic networks are another example of directed biological networks.

In recent years, attention has been given to the protein–protein interaction (PPI) networks of various organisms. In PPI networks, nodes represent proteins and edges depict physical or genetic interactions between them. The collection of all protein–protein interactions for one organism is identified by the term interactome. Systematic yeast two-hybrid (Y2H) and mass spectrometry-based biochemical affinity co-purification approaches have presumably revealed the dominant part of the yeast interactome. The human interactome, however, remains mostly unknown. PPI networks are usually modelled as undirected graphs, in which nodes characterize proteins and edges represent interactions.

Each of these biological networks has been reconstructed differently. A PPI network is a collection of diverse types of interaction from experimentally validated physical interactions to computationally predicted interactions. Reconstruction of regulatory or signalling networks directly from experimental data using statistical or machine learning approaches has received significant attention in the last decade. Suppose that, broadly speaking, we have a set of measurements from a system of interest, such as node attributes or binary indicators of certain existing edges but not others, or both of them. Then the aim of network inference methods is to select an appropriate network among all other possible networks that best captures the underlying state of the system. The evaluation criteria are defined normally based on consistency with the information in the data as well as any other prior information. Approaches employed for this task include Bayesian networks, auto-regressive models, correlation-based, mutual-information-based models, clustering techniques and differential equation models [80–86]. Most of these methods use perturbation to validate the reconstructed network. This can be performed by removing a node, gene, protein or metabolite from a system or by preventing interaction among particular ones of these.

3.4   Concluding Remarks

Despite its success, the purely topological approach possesses inherent limitations in the race for understanding cellular networks. In focusing on topology alone, we have neglected the fact that not all edges are created equal. In practice, the dynamical functionality of a complex network is probably affected not just by the binary pattern of who is connected to whom, but also by the nature of this connection, by its strength, and change over the time. Indeed, in a realistic biological network, several reactions are more transient than others – a feature that is overlooked by topology-based analyses. To summarize, biological networks are far more complex and diverse than can be understood purely using a graph model and topological analysis. They can change over time, have various types of links and different kinds of nodes, and even each node can be a complex, nonlinear, dynamical system by itself.

In the following chapters you will learn about new ways of studying such complex networks.

References

[1] Barabási AL, Oltvai ZN. Network biology: understanding the cell's functional organization. *Nat Rev Genet*. 2004;5:101–113.

[2] Sharma P, Ahmed HA, Roy S, Bhattacharyya DK. Unsupervised methods for finding protein complexes from PPI networks. *Netw Model Anal Heal Informatics Bioinforma*. 2015;4. doi:10.1007/s13721-015-0080-7

[3] Zenil H, Kiani NA, Tegnér J. Methods of information theory and algorithmic complexity for network biology. *Semin Cell Dev Biol*. 2016;51:32–43. doi:10.1016/j.semcdb.2016.01.011

[4] Kolaczyk ECG. *Visualizing Network Data: Statistical Analysis of Network Data with R*. New York: Springer, 2009.

[5] Broido AD, Clauset A. Scale-free networks are rare. *Nat Commun*. 2019;10. doi:10.1038/s41467-019-08746-5

[6] Holme P. Rare and everywhere: perspectives on scale-free networks. *Nat Commun*. 2019;10.

[7] Yu H, Greenbaum D, Lu HX, Zhu X, Gerstein M. Genomic analysis of essentiality within protein networks. *Trends Genet*. 2004;20:227–231.

[8] Han JDJ, Berlin N, Hao T, et al. Evidence for dynamically organized modularity in the yeast protein–protein interaction network. *Nature*. 2004;430:88–93. doi:10.1038/nature02555

[9] Luscombe NM, Babu MM, Yu H, et al. Genomic analysis of regulatory network dynamics reveals large topological changes. *Nature*. 2004;431:308–312. doi:10.1038/nature02782

[10] Ekman D, Light S, Björklund AK, Elofsson A. What properties characterize the hub proteins of the protein–protein interaction network of *Saccharomyces cerevisiae*? *Genome Biol*. 2006;7. doi:10.1186/gb-2006-7-6-r45

[11] Maslov S, Sneppen K. Specificity and stability in topology of protein networks. *Science*. 2002;296:910–913. doi:10.1126/science.1065103

[12] Han JDJ, Dupuy D, Bertin N, Cusick ME, Vidal M. Effect of sampling on topology predictions of protein–protein interaction networks. *Nat Biotechnol*. 2005;23:839–844.

[13] Kuchaiev O, Pržulj N. Learning the structure of protein–protein interaction networks. In: *Pacific Symposium on Biocomputing*, 2009, pp 39–50.

[14] Watts DJ, Strogatz SH. Collective dynamics of 'small-world' networks. In: *The Structure and Dynamics of Networks*. Princeton, NJ: Princeton University Press, 2011, pp 301–303.

[15] Zhang Z, Zhang J. A big world inside small-world networks. *PLoS One*. 2009;4:e5686. doi:10.1371/journal.pone.0005686

[16] Xu K, Bezakova I, Bunimovich L, Yi SV. Path lengths in protein–protein interaction networks and biological complexity. *Proteomics*. 2011;11:1857–1867. doi:10.1002/pmic.201000684

[17] Yook SH, Oltvai ZN, Barabási AL. Functional and topological characterization of protein interaction networks. *Proteomics*. 2004;4:928–942. doi:10.1002/pmic.200300636

[18] Gursoy A, Keskin O, Nussinov R. Topological properties of protein interaction networks from a structural perspective. In: *Biochemical Society Transactions*, 2008, pp 1398–1403.

[19] Sabidussi G. The centrality index of a graph. *Psychometrika*. 1966;31:581–603. doi:10.1007/BF02289527

[20] Freeman LC. A set of measures of centrality based on betweenness. *Sociometry*. 1977;40:35. doi:10.2307/3033543

[21] Freeman LC. Centrality in social networks conceptual clarification. *Soc Networks*. 1978;1:215–239. doi:10.1016/0378-8733(78)90021-7

[22] Lazega E, Van Duijn M. Position in formal structure, personal characteristics and choices of advisors in a law firm: a logistic regression model for dyadic network data. *Soc Networks*. 1997;19:375–397. doi:10.1016/S0378-8733(97)00006-3

[23] Bonacich P. Factoring and weighting approaches to status scores and clique identification. *J Math Sociol*. 1972;2:113–120. doi:10.1080/0022250X.1972.9989806

[24] Katz L. A new status index derived from sociometric analysis. *Psychometrika*. 1953;18:39–43. doi:10.1007/BF02289026

[25] Berman A, Plemmons RJ. *Nonnegative Matrices in the Mathematical Sciences*. Philadelphia, PA: Society for Industrial and Applied Mathematics, 1994.

[26] Hwang D-U, Boccaletti S, Moreno Y, López-Ruiz R. Thresholds for epidemic outbreaks in finite scale-free networks. *Math Biosci Eng*. 2005;2:317–327. doi:10.3934/mbe.2005.2.317

[27] Tong AHY, Evangelista M, Parsons AB, et al. Systematic genetic analysis with ordered arrays of yeast deletion mutants. *Science*. 2001;294:2364–2368. doi:10.1126/science.1065810

[28] Milo R, Kashtan N, Itzkovitz S, Newman MEJ, Alon U. On the uniform generation of random graphs with prescribed degree sequences. *arXiv:cond-mat/0312028*, 2003.

[29] Ma X, Gao L. Biological network analysis: insights into structure and functions. *Brief Funct Genomics*. 2012;11:434–442. doi:10.1093/bfgp/els045

[30] Qian J, Hintze A, Adami C. Colored motifs reveal computational building blocks in the *C elegans* brain. *PLoS One*. 2011;6. doi:10.1371/journal.pone .0017013

[31] Alon U. Network motifs: theory and experimental approaches. *Nat Rev Genet*. 2007;8:450–461.

[32] Vázquez A, Dobrin R, Sergi D, et al. The topological relationship between the large-scale attributes and local interaction patterns of complex networks. *Proc Natl Acad Sci USA*. 2004;101:17940–17945. doi:10.1073/pnas.0406024101

[33] Zenil H, Kiani NA, Tegnér J. Quantifying loss of information in network-based dimensionality reduction techniques. *J Complex Netw*. 2016;4. doi:10.1093/ comnet/cnv025

[34] Ravasz E, Barabási AL. Hierarchical organization in complex networks. *Phys Rev E Stat Physics Plasmas Fluids Relat Interdiscip Top*. 2003;67:7. doi:10.1103/PhysRevE.67.026112

[35] Barabási A-L, Ravasz E, Oltvai Z. Hierarchical organization of modularity in complex networks. *Stat Mech Complex Netw*. 2003;625:46–65.

[36] Papadopoulos S, Kompatsiaris Y, Vakali A, Spyridonos P. Community detection in social media performance and application considerations. *Data Min Knowl Discov*. 2012;24:515–554. doi:10.1007/s10618-011-0224-z

[37] Zhang P. Evaluating accuracy of community detection using the relative normalized mutual information. *J Stat Mech Theory Exp*. 2015. doi:10.1088/ 1742-5468/2015/11/P11006

[38] Yang Z, Algesheimer R, Tessone CJ. A comparative analysis of community detection algorithms on artificial networks. *Sci Rep*. 2016;6. doi:10.1038/srep30750

[39] Peel L, Larremore DB, Clauset A. The ground truth about metadata and community detection in networks. *Sci Adv.*, 2017;3. doi:10.1126/sciadv.1602548

[40] Radicchi F, Castellano C, Cecconi F, Loreto V, Paris D. Defining and identifying communities in networks. *Proc Natl Acad Sci USA*. 2004;101:2658–2663. doi:10.1073/pnas.0400054101

[41] Fortunato S, Barthélemy M. Resolution limit in community detection. *Proc Natl Acad Sci USA*. 2007;104. doi:10.1073/pnas.0605965104

[42] Brandes U, Delling D, Gaertler M, et al. On modularity clustering. *IEEE Trans Knowl Data Eng*. 2008;20:172–188. doi:10.1109/TKDE.2007.190689

[43] Yang J, Leskovec J. Defining and evaluating network communities based on ground-truth. *Knowl Inf Syst*. 2015;42:181–213. doi:10.1007/s10115-013-0693-z

[44] Jin D, He D, Hu Q, Baquero C, Yang B. Extending a configuration model to find communities in complex networks. *J Stat Mech Theory Exp* 2013. doi:10.1088/1742-5468/2013/09/P09013

[45] Ibrahim ZM, Ngom A. The relative vertex clustering value: a new criterion for the fast discovery of functional modules in protein interaction networks. *BMC Bioinformatics*. 2015;16. doi:10.1186/1471-2105-16-S4-S3

[46] Srinivas S, Rajendran C. Community detection and influential node identification in complex networks using mathematical programming. *Expert Syst Appl*. 2019;135:296–312. doi:10.1016/j.eswa.2019.05.059

[47] Delvenne JC, Yaliraki SN, Barahon M. Stability of graph communities across time scales. *Proc Natl Acad Sci USA*. 2010;107:12755–12760. doi:10.1073/pnas.0903215107

[48] Fortunato S. Community detection in graphs. *Phys Rep*. 2010;486:75–174.

[49] Fortunato S, Hric D. Community detection in networks: A user guide. *Phys Rep*. 2016;659:1–44.

[50] Lancichinetti A, Fortunato S. Community detection algorithms: a comparative analysis. *Phys Rev E Stat Nonlinear Soft Matter Phys*. 2009;80. doi:10.1103/PhysRevE.80.056117

[51] Raghavan UN, Albert R, Kumara S. Near linear time algorithm to detect community structures in large-scale networks. *Phys Rev E Stat Nonlinear Soft Matter Phys*. 2007;76. doi:10.1103/PhysRevE.76.036106

[52] Dahlin J, Svenson P. Ensemble approaches for improving community detection methods. *arXiv:1309.0242*, 2013.

[53] Girvan M, Newman MEJ. Community structure in social and biological networks. *Proc Natl Acad Sci USA*. 2002;99:7821–7826. doi:10.1073/pnas.122653799

[54] Clauset A, Newman MEJ, Moore C. Finding community structure in very large networks. *Phys Rev E Stat Physics Plasmas Fluids Relat Interdiscip Top*. 2004;70:6. doi:10.1103/PhysRevE.70.066111

[55] Blondel VD, Guillaume JL, Lambiotte R, Lefebvre E. Fast unfolding of communities in large networks. *J Stat Mech Theory Exp* 2008. doi:10.1088/1742-5468/2008/10/P10008

[56] Sobolevsky S, Campari R, Belyi A, Ratti C. General optimization technique for high-quality community detection in complex networks. *Phys Rev E Stat Nonlinear Soft Matter Phys.* 2014;90. doi:10.1103/PhysRevE.90.012811

[57] Reichardt J, Bornholdt S. Statistical mechanics of community detection. *Phys Rev E Stat Nonlinear Soft Matter Phys.* 2006;74. doi:10.1103/PhysRevE.74.016110

[58] Guimerà R, Sales-Pardo M, Amaral LAN. Modularity from fluctuations in random graphs and complex networks. *Phys Rev E Stat Physics Plasmas Fluids Relat Interdiscip Top.* 2004;70:4. doi:10.1103/PhysRevE.70.025101

[59] Newman MEJ. Finding community structure in networks using the eigenvectors of matrices. *Phys Rev E Stat Nonlinear Soft Matter Phys.* 2006;74. doi:10.1103/PhysRevE.74.036104

[60] Xie J, Szymanski BK. Community detection using a neighborhood strength driven label propagation algorithm. *arXiv:1105.3264 [cs.SI]*, 2011.

[61] Pons P, Latapy M. Computing communities in large networks using random walks. *J Graph Algorithms Appl.* 2006;10:191–218. doi:10.7155/jgaa.00124

[62] Rosvall M, Bergstrom CT. An information-theoretic framework for resolving community structure in complex networks. *Proc Natl Acad Sci USA.* 2007;104:7327–7331. doi:10.1073/pnas.0611034104

[63] Rosvall M, Axelsson D, Bergstrom CT. The map equation. *Eur Phys J Spec Top.* 2009;178:13–23. doi:10.1140/epjst/e2010-01179-1

[64] Erdös P, Rényi A. On random graphs. I. *Publ Math.* 1959;6:290–297.

[65] Erdös P, Rényi A. On the evolution of random graphs. Magy Tudományos Akadémia Mat Kut Intézetének Közleményei [*Publ Math Inst Hungarian Acad Sci*]. 1960;5:17–61.

[66] Gilbert EN. Random graphs. *Ann Math Stat.* 1959;30:1141–1144. doi:10.1214/aoms/1177706098

[67] Bollobas B, Erdös P. Cliques in random graphs. *Math Proc Cambridge Philos Soc.* 1976;80:419–427. doi:10.1017/S0305004100053056

[68] Bollobas B. *Random Graphs.* Cambridge: Cambridge University Press, 2001.

[69] Bagrow JP. Evaluating local community methods in networks. *J Stat Mech Theory Exp.* 2008. doi:10.1088/1742-5468/2008/05/P05001

[70] Newman MEJ. The structure and function of complex networks. *SIAM Rev.* 2003;45:167–256.

[71] Albert R, Barabási AL. Statistical mechanics of complex networks. *Rev Mod Phys.* 2002;74:47–97. doi:10.1103/RevModPhys.74.47

[72] Watts DJ. *Small Worlds? The Dynamics of Networks Between Order and Randomness.* Princeton, NJ: Princeton University Press, 1999.

[73] Milgram S. The small world problem. *Psychol Today.* 1967;1:60–67.

[74] De Solla Price DJ. Networks of scientific papers. *Science.* 1965;149:510. doi:10.1126/science.149.3683.510

[75] Sheridan P, Onodera T. A preferential attachment paradox: how preferential attachment combines with growth to produce networks with log-normal in-degree distributions. *Sci Rep.* 2018;8. doi:10.1038/s41598-018-21133-2

[76] Price DDS. A general theory of bibliometric and other cumulative advantage processes. *J Am Soc Inf Sci.* 1976;27:292–306. doi:10.1002/asi.4630270505

[77] Albert R, Barabási A-L. Statistical mechanics of complex networks. *Rev Mod Phys.* 2002;74:47–97. doi:10.1103/RevModPhys.74.47

[78] Chao S-Y. Graph theory and analysis of biological data in computational biology. In: *Advanced Technologies.* London: InTech, 2009.

[79] Xia Y, Yu H, Jansen R, et al. Analyzing cellular biochemistry in terms of molecular networks. *Annu Rev Biochem.* 2004;73:1051–1087. doi:10.1146/annurev.biochem.73.011303.073950

[80] Kiani NA, Kaderali L. Dynamic probabilistic threshold networks to infer signaling pathways from time-course perturbation data. *BMC Bioinformatics.* 2014;15:250. doi:10.1186/1471-2105-15-250

[81] Marbach D, Costello JC, Kuffner R, et al. Wisdom of crowds for robust gene network inference. *Nat Methods.* 2012;9:796–804. doi:10.1038/nmeth.2016

[82] Newman MEJ, Clauset A. Structure and inference in annotated networks. *Nat Commun.* 2016;7. doi:10.1038/ncomms11863

[83] Deng Y, Zenil H, Tegnér J, Kiani NA. HiDi: an efficient reverse engineering schema for large-scale dynamic regulatory network reconstruction using adaptive differentiation. *Bioinformatics.* 2017;33. doi:10.1093/bioinformatics/btx501

[84] Margolin AA, Nemenman I, Basso K, et al. ARACNE: an algorithm for the reconstruction of gene regulatory networks in a mammalian cellular context. *BMC Bioinformatics.* 2006;7. doi:10.1186/1471-2105-7-S1-S7

[85] Khatamian A, Paull EO, Califano A, Yu J. SJARACNe: a scalable software tool for gene network reverse engineering from big data. *Bioinformatics.* 2019;35:2165–2166. doi:10.1093/bioinformatics/bty907

[86] He J, Zhou Z, Reed M, Califano A. Accelerated parallel algorithm for gene network reverse engineering. *BMC Syst Biol.* 2017;11:83. doi:10.1186/s12918-017-0458-5

[87] Guo W, Calixto CPG, Tzioutziou N, et al. Evaluation and improvement of the regulatory inference for large co-expression networks with limited sample size. *BMC Syst Biol.* 2017;11. doi:10.1186/s12918-017-0440-2

4 Networks from an Information-Theoretic and Algorithmic Complexity Perspective

Hector Zenil and Narsis A. Kiani

4.1 Introduction

On the one hand, the power of a computational model able to implement a complexity or information-theoretic measure, as shown in [1], is key to understanding its capabilities and limitations. For example, popular implementations of lossless compression algorithms such as LZW can be implemented using finite state automata (FSA) [2], which means they cannot reach the computational power needed to characterize algorithmic features – that is, the power of compression of algorithmic complexity that requires the full power of a universal Turing machine. FSAs will only be able to capture statistical properties and some very simple algorithmic features at the level of regular languages. Nevertheless, 'lossless compression' can be as general as necessary, all the way up to the strongest level in a computational hierarchy. In reality, however, popular compression algorithms occupy the weakest level, alongside, or slightly above, the power of certain other computable measures such as Shannon entropy [1].

It is very clear that popular lossless compression algorithms are limited by design, as their aim is to look for statistical regularities (i.e. repetitions) within a sliding window of variable length in order to compile a dictionary in which the most frequently repeated long segments are replaced by shorter codes. When such a window length is unbounded, algorithms are said to be optimal and even universal, as they can approach exact values of the entropy rate at the limit, but they are not optimal or universal in the algorithmic sense. Popular lossless compression algorithms such as LZW then implement a look-up table based on a dictionary compiled by assigning the longest repetitions to the shorter codes, something that an algorithmic measure would naturally do, beyond statistical regularities. Other lossless compression algorithms are directly based on the number of repetitions or Shannon entropy, with most, if not all, strategies being very similar in essence, that is statistical.

On the other hand, network theory is a central topic in computational systems, such as in molecular biology, serving as a framework within which to understand and reconstruct relations among biological components. For example, constructing networks from a gene expression dataset provides a set of possible hypotheses explaining connections among genes, knowledge that is vital to advancing our understanding of living organisms as systems.

Over the last decade, network theory has become a unifying language in biology, giving rise to whole new areas of research in computational systems biology. Gene networks are conceptual models of genetic regulation, in which each gene is considered to be directly affected by a number of other genes, usually represented by a directed graph.

Graphs are an important tool for the mathematical analysis of many systems, from the interactions of chemical agents, cells and genes, to ecological networks. Yet we are only just beginning to understand the significance of network biology.

Classical information theory has for some time been applied to networks, but as we will illustrate, Shannon entropy, like other computable measures (i.e. one that is a total function returning an output for every input in finite time), is not invariant to changes of object description [3].

4.1.1   Graph Notation and Complex Networks

A graph G is *labelled* when the vertices are distinguished by names such as $u_1, u_2, \ldots u_n$ with $n = |V(G)|$. Graphs G and H are said to be *isomorphic* if there is a bijection between the vertex sets of G and H, $\lambda : V(G) \rightarrow V(H)$ such that any two vertices u and $v \in G$ are adjacent in G if and only if $\lambda(u)$ and $\lambda(v)$ are adjacent in H. When G and H are the same graph, the bijection is referred to as an *automorphism* of G. The adjacency matrix of a graph is not an invariant under *graph relabellings*. Figure 4.1 shows two adjacency matrices for isomorphic graphs. $A(G)$ denotes the adjacency matrix of G. $|V(G)|$ and $|E(G)|$ denote the vertex and edge count of G.

One of the most basic properties of graphs is the number of links per node. When all nodes have the same number of links, the graph is said to be *regular*. The *degree* of a node v, denoted by $d(v)$, is the number of (incoming and outgoing) links to other nodes. We will also say that a graph is *planar* if it can be drawn in a plane without its edges crossing. Planarity is an interesting property because only planar graphs have *duals*. A *dual graph* of a planar graph G is a graph that has a vertex corresponding to each face of G, and an edge joining two neighbouring faces for each edge in G.

The graph spectrum of a graph is the set of eigenvalues of its adjacency matrix. Two graphs are called *co-spectral* if the adjacency matrices of the graphs have equal multisets of eigenvalues.

A *canonical form* of G is a labelled graph $Canon(G)$ that is isomorphic to G, such that every graph that is isomorphic to G has the same canonical form as G. An advantage of $Canon(G)$ is that unlike $A(G)$, $A(Canon(G))$ is a graph invariant of $Canon(G)$ [4].

A popular type of graph that has been studied is the so-called *Erdős-Rényi* [5, 6] (ER) graph, in which vertices are randomly and independently connected by links with a fixed probability (also called *edge density*) (see Figure 4.2 for a comparison

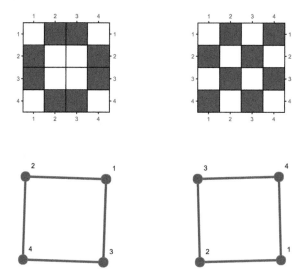

Figure 4.1 Isomorphic graphs with two different adjacency matrix representations which translate into graph relabellings, illustrating that the adjacency matrix is not an invariant of an unlabelled graph. However, similar graphs have adjacency matrices with similar algorithmic information content, as proven in [7].

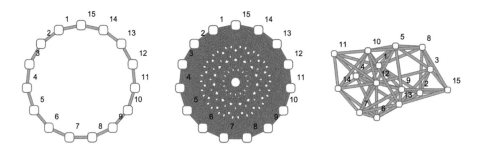

Figure 4.2 Examples of simple versus random graphs characterized by algorithmic complexity (K). $K(G) \sim \log_2 |V(G)| = \log_2 15 \sim 3.9$ bits when a graph is simple and is highly compressible. In contrast, a random graph (right) with the same number of nodes and number of links requires more information to be specified, because there is no simple rule connecting the nodes and therefore $K(G) \sim |E(G)| = 15$ in bits, i.e. the ends of each edge have to be specified (so a tighter bound would be $2|E(G)| \sim 30$ for an ER graph of edge density ~ 0.5.

between a regular and a random graph of the same size). The probability of vertices being connected is referred to as the *edge probability*. The main characteristic of random graphs is that all nodes have roughly the same number of links, equal to the average number of links per node. An ER graph $G(n, p)$ is a graph of size n constructed by connecting nodes randomly with probability p independent of every other edge. Usually ER graphs are assumed to be non-recursive (i.e. truly random), but ER graphs can be constructed recursively with, for example, pseudorandom algorithms. Here we will assume that ER graphs are non-recursive, as theoretical comparisons and bounds hold only in the non-recursive case. For numerical estimations, however, we use a pseudorandom edge connection algorithm, in keeping with common practice.

The so-called *small-world* graph describes the phenomenon of many empirical networks in which most vertices are separated by a relatively small number of edges. A network is considered a *small-world* graph G if the average graph distance D grows no faster than the log of the number of nodes: $D \sim \log|V(G)|$. Many networks are *scale-free*, meaning that their degrees are size independent, in the sense that the empirical degree distribution is independent of the size of the graph up to a logarithmic term. That is, the proportion of vertices with degree k is proportional to γk^τ for some $\tau > 1$ and constant γ. In other words, many empirical networks display a power-law degree distribution.

4.1.2   Classical Information Theory

Information theory is a field that specifies fundamental limits on signal processing, such as communicating, storing and compressing data. Central to information theory is the concept of Shannon's information entropy, which quantifies the average number of bits needed to store or communicate a message. Shannon's entropy determines that one cannot store (and therefore communicate) a symbol with n different symbols in less than $\log(n)$ bits. In this sense, Shannon's entropy determines a lower limit below which no message can be further compressed, even in principle. Another application (or interpretation) of Shannon's information theory is as a measure for quantifying the *uncertainty* involved in predicting the value of a random variable. For example, specifying the outcome of a fair coin flip (two equally likely outcomes) requires one bit at a time, because the results are independent and therefore each result conveys maximum entropy. Things begin to get interesting when the coin is not fair. If one considers a coin with heads on both obverse and reverse, then the tossing experiment always results in heads, and the message will always be 1 with full certainty.

For an ensemble $X(R, p(x_i))$, where R is the set of possible outcomes (the random variable), $n = |R|$ and $p(x_i)$ is the probability of an outcome in R. The Shannon information content or entropy of X is then given by

$$H(X) = -\sum_{i=1}^{n} p(x_i) \log_2 p(x_i) \tag{4.1}$$

which implies that to calculate $H(X)$ one has to know or assume the mass distribution probability of ensemble X. One caveat regarding Shannon's entropy is that one is forced to make an arbitrary choice regarding granularity. Take, for example, the bit string 01010101010101. The Shannon entropy of the string at the level of single bits is maximal, as there are the same number of 1s and 0s, but the string is clearly regular when two-bit blocks are taken as basic units, in which instance the string has minimal complexity because it contains only one symbol (01) from among the four possible symbols (00,01,10,11). One way to overcome this problem is to take into consideration all possible 'granularities' (we call this *block entropy*), from length 1 to n, where n is the length of the sequence. This measure is related to what's also called *predictive information* or *excess entropy* (the differences among the entropies for consecutive block sizes).

To proceed by means of block entropy is computationally expensive, as compared to fixing the block size at n, as it entails producing all possible overlapping $\binom{i}{n}$ substrings for all $i \in \{1, \ldots, n\}$.

4.2   Graph Entropy

One of the major challenges in modern physics is to provide proper and suitable representations of network systems for use in fields ranging from physics [8] to chemistry [9]. A common problem is the description of order parameters with which to characterize the *complexity of a network*. Graph complexity has traditionally been characterized using graph-theoretic measures such as degree distribution, clustering coefficient, edge density, and community or modular structure.

More recently, networks have also been characterized using classical information theory. One problem in this area is the interdependence of many graph-theoretic properties, which makes measures more sophisticated than single-property measurements [10] difficult to come by. The standard way to address this is to generate graphs that have a certain specific property while being random in all other respects, in order to check whether or not the property in question is typical among an ensemble of graphs with otherwise seemingly different properties.

Approaches using measures based upon Shannon entropy which claim to quantify the information content of a network [11] as an indication of its 'typicality' are based on an assumption of associated ensembles provided by the entropy evaluation: the more random the more typical. The claim is that one can construct a 'null model' that captures some aspects of a network (e.g. graphs that have the same degree distribution) and see how different the network is to the null model as regards particular features, such as clustering coefficient, graph distance or other features of interest. The procedure aims at producing an intuition of an ensemble of graphs that are assumed to have been sampled uniformly at random from the set of all graphs with the same property, in order to determine if such a property occurs with high or low probability. If the graph is not significantly different, statistically, from the null model, then the graph is said to be as 'simple' as the null model; otherwise, the measure is said to be a lower bound on the 'complexity' of the graph as an indication of its random as opposed to causal nature.

While the application of entropy to graph degree distributions has been relatively more common, the same entropy has also been applied to other graph features, such as functions of their adjacency matrices [12], and to distance and Laplacian matrices [13].

Even more recently, Shannon entropy on adjacency matrices was used to attempt the discovery of CRISPR regions in an interesting transformation of DNA sequences into graphs [14]. A survey contrasting adjacency matrix-based (walk) entropies and other entropies (e.g. on degree sequence) is offered in [12]. It finds that adjacency-based ones are more robust vis-a-vis graph size and are correlated to graph algebraic properties, as these are also based on the adjacency matrix (e.g. graph spectrum).

In estimating the complexity of objects, in particular of graphs, it is common practice to rely on graph- and information-theoretic measures. Here, using integer sequences with properties such as Borel normality, we explain how these mea-sures are not independent of the way in which an object, such as a graph, can be described or observed. From observations that can reconstruct the same graph and are therefore essentially translations of the same description, we will see that when applying a computable measure such as Shannon entropy, not only is it necessary to pre-select a feature of interest where there is one, and to make an arbitrary selection where there is not, but also that more general properties, such as the causal likelihood of a graph as a measure (opposed to randomness) can be largely misrepresented by computable measures such as entropy and entropy rate. We introduce recursive and non-recursive (uncomputable) graphs and graph constructions based on these integer sequences, whose different lossless descriptions have disparate Entropy values, thereby enabling the study and exploration of a measure's range of applications and demonstrating the weaknesses of computable measures of complexity.

Just as for strings, Shannon entropy can also be applied to the node degree sequence of a graph. One of the most popular applications of Entropy is to graph degree distribution, as first suggested and introduced by [15]. Similar approaches have been adopted in areas such as chemical graph theory and computational systems biology [16] as functions of layered graph degree distribution under certain layered coarse-graining operations (sphere covers), leading to the hierarchical application of entropy, a version of graph traversal entropy rate. In chemistry, for example, Shannon entropy over a function of degree sequence has been used as a profiling tool to characterize – so it is claimed – molecular complexity.

The Shannon entropy of an unlabelled network characterized by its degree distri-bution can be described by the same formula for Shannon entropy where the random variable is a degree distribution. The chief advantage of so doing is that it is invariant to relabellings. This also means that the degree distribution is not a lossless represen-tation of a labelled network (but rather of its isomorphic group), and is an interesting entropic measure, but one that can only be used when the node labels are not relevant. For example, in a causal recursive network, the node labels may represent time events in a sequence that has a meaning captured in the network labelling, in which case the degree distribution sequence (where no agreement has been reached on the order of the elements in the distribution sequence, which is therefore disordered) cannot be used to reconstruct the original network represented by the unlabelled version or the isomorphism group of the labelled networks.

It is also clear that the concept of entropy rate cannot be applied to the degree distribution, because the node degree sequence has no particular order, or any order is meaningless because any label numbering will be arbitrary. This also means that Shannon entropy is not invariant to the language description of a network, especially as a labelled or an unlabelled network, except for clearly extreme cases (e.g. fully disconnected and completely connected networks, both of which have flat degree distributions and therefore the lowest Shannon entropy for degree sequence and adjacency matrix).

Another review and comparison among these entropic measures is provided in [17] in connection with applications to drug and chemical structures.

4.2.1   The Fragility of Entropy and Computable Measures

Traditional statistical and computable measures such as Shannon entropy can easily be proven not to be robust and to require arbitrary choices at multiple levels. Dependent on underlying mass distributions, and mostly quantifying how removed the assumptions of the premises are from reality, Shannon entropy is very limited in dealing with complexity, information content, and ultimately, causation.

The Shannon entropy (or simply entropy) of a graph G can defined by

$$H(A(G)) = -\sum_{i=1}^{n} P(A(x_i)) \log_2 P(A(x_i))$$ (4.2)

where G is the random variable with n possible outcomes (all possible adjacency matrices of size $|V(G)|$). For example, a completely disconnected graph G with all adjacency matrix entries equal to zero has entropy $H(A(G)) = 0$, because the number of different symbols in the adjacency matrix is 1. However, if a different number of 1s and 0s occur in $A(G)$, then $H(A(G)) \neq 0$. In general, we will use block entropy in order to detect more graph regularities (through the adjacency matrix) at a greater resolution. But for block entropy there is an extra factor to be taken into account. The unlabelled calculation of the block entropy (not relevant for 1-bit entropy) of a graph has to take into consideration all possible adjacency matrix representations for all possible labellings. Therefore, the block entropy of a graph is given by:

$$H(G) = \min\{H(A(g_L))|G_L \in L(G)\}$$ (4.3)

where $L(G)$ is the group of all possible labellings of G.

Other entropic measures of network elements are possible and have been proposed before, but they all require one to focus on a particular graph element (adjacency matrix, degree sequence, number of bifurcations), and they do not all converge, meaning that Shannon entropy is not invariant to different descriptions of the same object, thus failing, unlike algorithmic complexity, to characterize any general or universal property of a graph or network [18]. Indeed, in [18] we introduced a graph that is generated recursively by a small computer program of (small) fixed length, yet when looking at its degree sequence it tends to maximal entropy and when looking at the adjacency matrix it tends to zero entropy at the limit, thus displaying divergent values for the same object when considering different mass probability distributions – when assuming the uniform distribution to characterize an underlying ensemble comprising all possible adjacency matrices of increasing size, or when assuming all possible degree sequences in the face of a total lack of knowledge of the deterministic nature of the graph in question.

More recently, we have also proposed a refinement of the so-called principle of maximum entropy, or Maxent, based on algorithmic complexity. We have shown, for example, that not all ER networks are random [19], and that our methods can to some extent tell apart random from pseudorandom ER networks, questioning the ability of classical Maxent to compare objects against their most randomized versions.

4.3   Graph Algorithmic Complexity

In contrast, or complementary to computable measures, we have introduced methods
to approximate the algorithmic complexity of a graph with interesting results [7, 20,
21]. For example, in [7] correlations were reported among algebraic and topological
properties of synthetic and biological networks by means of algorithmic complexity, and an application to classify networks by type was developed in [20]. Together
with [7] and [20], the methods introduced represented a novel view and constitute
a formal approach to graph complexity, while providing a new set of tools for the
analysis of the local and global structures of networks.

Formally, the algorithmic complexity [22, 23] of a string s is evaluated as follows:

$$K(s) = \min\{|p| : U(p) = s\} \tag{4.4}$$

That is, the length (in bits) of the shortest program p that when running on a universal
Turing machine U outputs s upon halting.

A universal Turing machine U is an abstraction of a general-purpose computer
that can be programmed to reproduce any computable object, such as a string or a
network (e.g. the elements of an adjacency matrix). By the *invariance theorem* [24, 25],
K_U only depends on U up to a constant, so as is conventional, the U subscript can
be dropped. Formally, $\exists \gamma$ such that $|K_U(s) - K_{U'}(s)| < \gamma$ where γ is a constant inde-
pendent of U and U'. Because everything in the theory of Kolmogorov complexity is
meant to be asymptotic, this invariance theorem means that the longer the string s the
closer the algorithmic complexity evaluations, even for different Turing machines U
and U', and at the limit (for $|s| \to \infty$) the evaluations will coincide.

Due to its power, K comes burdened with a technical inconvenience (formally
called *semi-computability*) and it has been proven that no effective algorithm exists
which takes a string s as input and produces the exact integer $K(s)$ as output [22, 23].
This is related to a common problem in computer science known as the undecidability
of the halting problem [26] – referring to the ability to know whether or not a computation will eventually stop.

Despite the inconvenience, K can be effectively approximated by using, for example, compression algorithms. Algorithmic complexity can alternatively be understood
in terms of uncompressibility. If an object, such as a biological network, is highly
compressible, then K is small and the object is said to be non-random. However, if
the object is uncompressible then it is considered algorithmically random.

A compression ratio, also related to the *randomness deficiency* of a network or how
removed an object is from maximum algorithmic randomness, will be defined by
$C(G) = Comp(G)/|A(G)|$, where $Comp(G)$ is the compressed length in bits of the adjacency matrix G of a network using a *lossless* compression algorithm (e.g. Compress),
and $|A(G)|$ is the size of the adjacency matrix measured by taking the dimensions
of the array and multiplying its values (e.g., if the adjacency matrix is 10×10, then
$|A(G)| = 100$). It is worth mentioning that compressibility is a sufficient test for non-
randomness. This means that it does not matter which lossless compression algorithm
is used – if an object can be compressed then it is a valid upper bound of its algorithmic
complexity. This in turn means that the choice of lossless compression algorithm is
not very important, because one can test them one by one and always retain the

best compression as an approximation to K. A lossless compression algorithm is an algorithm that includes a decompression algorithm that retrieves the exact original object, without any loss of information when decompressed. The closer $C(G)$ is to 1 the less compressible, and the closer to 0 the more compressible.

4.3.1 Algorithmic Probability

The algorithmic probability [27–29] of a string s, denoted by $AP(s)$ provides the probability that a valid random program p written in bits uniformly distributed produces the string s when run on a universal (prefix-free[1]) Turing machine U. Formally,

$$AP(s) = \sum_{p:U(p)=s} 1/2^{|p|} \qquad (4.5)$$

That is, the sum over all the programs p for which a universal Turing machine U outputs s and halts.

Algorithmic probability and algorithmic complexity K are formally (inversely) related by the so-called algorithmic coding theorem [24, 30]:

$$|-\log_2 AP(s) - K(s)| < \mathcal{O}(1) \qquad (4.6)$$

where $\mathcal{O}(1)$ is an additive value independent of s. This coding theorem implies that one can estimate the algorithmic complexity of a string from its frequency.

Assuming Borel's absolute normality of a mathematical constant such as π whose digits appear randomly distributed, and with no knowledge of the deterministic source and nature of π as produced by short mathematical formulae, the Shannon entropy rate (thus assuming the uniform distribution along all integer sequences of N digits) of the N first digits of π, in any base, would suggest maximum randomness at the limit. However, without access to or assumptions of probability distribution, approximations to algorithmic probability would assign π high probability and, thus, lowest complexity by the coding theorem, as has been done in [31–35].

4.3.2 Approximations to Graph Algorithmic Complexity

As shown in [7], estimations of algorithmic complexity are able to distinguish complex from random networks (of the same size, or growing asymptotically), which are both in turn distinguished from regular graphs (also of the same size). K calculated by the block decomposition method (BDM) assigns low algorithmic complexity to regular graphs, medium complexity to complex networks following Watts–Strogatz or Barabási–Albert algorithms, and higher algorithmic complexity to random networks. That random graphs are the most algorithmically complex is clear from a theoretical point of view: nearly all long binary strings are algorithmically random, and so nearly all random unlabelled graphs are algorithmically random [36], where

---

[1] The group of valid programs forms a prefix-free set (no element is a prefix of any other, a property necessary to keep $0 < AP(s) < 1$.) For details see [24, 30].

algorithmic complexity is used to give a proof of the number of unlabelled graphs as a function of its randomness deficiency (how far it is from the maximum value of $K(G)$).

The *coding theorem method* (CTM) [31, 32] is rooted in the relation specified by algorithmic probability between frequency of production of a string from a random program and its algorithmic complexity (Eq. 4.6). It is also called the algorithmic *coding theorem*, to contrast it with another coding theorem in classical information theory. Essentially, it uses the fact that the more frequent a string (or object), the lower its algorithmic complexity; and strings of lower frequency have higher algorithmic complexity.

The way to implement a compression algorithm at the level of Turing machines, unlike popular compression algorithms based on Shannon entropy, is to go through all possible compression schemes, the equivalent of going through all possible programs that generate a piece of data, which is exactly what my CTM algorithm does.

In [7], numerical evidence was provided in support of the theoretical assumption that the algorithmic complexity of an unlabelled graph should not be far removed from that of any of its labelled versions. This is because there is a small computer program of fixed size that should determine the order of the labelling proportional to the size of the isomorphism group. Indeed, when the isomorphism group is large, the labelled networks have more equivalent descriptions given by the symmetries, and would therefore, according to algorithmic probability, be of lower algorithmic complexity.

4.3.3 Reconstructing K from Local Graph Algorithmic Patterns

The approach to determining the algorithmic complexity of a graph thus involves considering how often the adjacency matrix of a motif is generated by a random Turing machine on a two-dimensional array, also called a *termite* or *Langton's ant* [37]. We call this the *block decomposition method*, as introduced in [7] and [35], as it requires the partition of the adjacency matrix of a graph into smaller matrices, using which we can numerically calculate its algorithmic probability by running a large set of small two-dimensional deterministic Turing machines, and then – by applying the algorithmic coding theorem – its algorithmic complexity. Then the overall complexity of the original adjacency matrix is the sum of the complexity of its parts, albeit with a logarithmic penalization for repetition, given that n repetitions of the same object only add $\log n$ to its overall complexity, as one can simply describe a repetition in terms of the multiplicity of the first occurrence. More formally, the algorithmic complexity of a labelled graph G by means of BDM is defined as follows:

$$K_{BDM}(G,d) = \sum_{(r_u,n_u) \in A(G)_{d \times d}} \log_2(n_u) + K_m(r_u) \qquad (4.7)$$

where $K_m(r_u)$ is the approximation of the algorithmic complexity of the subarrays r_u arrived at by using the algorithmic coding theorem (Eq. 4.6), while $A(G)_{d \times d}$ represents the set with elements (r_u, n_u), obtained by decomposing the adjacency matrix of G into non-overlapping squares of size d by d. In each (r_u, n_u) pair, r_u is one such square and n_u its multiplicity (number of occurrences). From now on, $K_{BDM}(g, d = 4)$ will be denoted only by $K(G)$, but it should be taken as an approximation to $K(G)$ unless otherwise stated (e.g. when taking the theoretically true $K(G)$ value). Once CTM is calculated, BDM can be implemented as a look-up table, and hence runs efficiently in linear time for non-overlapping fixed size submatrices.

As with block entropy (see Section 4.3), the algorithmic complexity of a graph G is given by:

$$K'(G) = \min\{K(A(G_L))|G_L \in L(G)\} \tag{4.8}$$

where $L(G)$ is the group of all possible labellings of G and G_L a particular labelling. In fact, $K(G)$ provides a choice for graph canonization, taking the adjacency matrix of G with the lowest algorithmic complexity. Because it may not be unique, it can be combined with the smallest lexicographical representation when the adjacency matrix is concatenated by rows, as is traditionally done for graph canonization. By taking subarrays of the adjacency matrix we ensure that network motifs (over-represented graphs), used in biology and proven to classify superfamilies of networks [39], are taken into consideration in the BDM calculation. And indeed we were able to show that BDM alone classifies the same superfamilies of networks [20] that classical network motifs were able to identify.

4.3.4   Robustness of Algorithmic Graph Complexity

Despite the complexity of the calculation of unlabelled complexity K', regular graphs have been shown to have low K and random graphs have been shown to have high K estimations, with graphs with a larger set of automorphisms having lower K than graphs with a smaller set of automorphisms [7]. An important question is how accurate a labelled estimation of $K(G)$ is with respect to the unlabelled $K'(G)$, especially because in the general case the calculation of $K(G)$ is computationally cheap compared to $K'(G)$, which carries an exponential overhead. However, the difference $|K(G) - K'(G)|$ is bounded by a constant. Indeed, as first suggested in [33], there exists an algorithm α of fixed length $|\alpha|$ bits such that one can compute all $L(G)$ relabellings of G, even if by brute force (e.g. by producing all the indicated adjacency matrix row and column permutations). Therefore $|K(G) - K(G_L)| < |\alpha|$ for any relabelled graph G_L of G, or in other words, $K(G_L) = K'(G) + |\alpha|$, where $|\alpha|$ is independent of G. Notice, of course, that here the time complexity of α believed to not be in $\mathbf{P}$ is irrelevant; what is needed for the proof is that it exists and is therefore of finite size. We can therefore safely estimate the unlabelled $K'(G)$ by estimating a labelled $K(G_L)$ as an accurate asymptotic approximation. In fact, brute force is likely the shortest program description to produce all relabellings, and therefore the best choice to minimize α.

This result is relevant, first because it means one can accurately estimate $K_L(G)$ through $K(G)$ for any lossless representation of G up to an additive term. One problem is that this does not tell us the rate of convergence of $K(G)$ to $K_L(G)$. But numerical estimations show that the convergence is in practice fast. For example, the median of the BDM estimations of all the isomorphic graphs of the graph in Figure 4.1 is 31.7, with a standard deviation of 0.72. However, when generating a graph, the BDM median is 27.26 and the standard deviation 2.93, clearly indicating a statistical difference. But more importantly, the probability of a random graph having a large automorphism group count is low, as shown in [7], which is consistent with what we would expect of the algorithmic probability of a random graph – a low frequency of production as a result of running a *termite* Turing machine. And here and in [7] we have also shown that graphs and their dual and co-spectral versions have similar algorithmic complexity values as calculated by algorithmic probability (BDM), hence indicating that in practice the convergence guaranteed by the result in this section is fast.

4.3.5 $K(G)$ is Not a Graph Invariant but Highly Informative of G

$K(G)$ may be computationally cheap to approximate up to a bounded error that vanishes as a function of the size of the graph. However, $K(G)$ does not uniquely determine G. Indeed, two non-isomorphic graphs G and H can have $K(G) = K(H)$. In fact, the algorithmic coding theorem gives an estimation of how often this happens, and it is also related to a simple pigeonhole argument. Indeed, if G or H are algorithmically (Kolmogorov) random graphs, then the probability that $K(G) = K(H)$ grows exponentially. If G and H are complex, then their algorithmic probabilities $\sim 1/2^K(G)$ and $\sim 1/2^K(H)$ respectively are small and are in the tail of the algorithmic probability (the so-called *universal distribution* or Levin's *semi-measure*) distribution ranging across a very small interval of (maximal) algorithmic complexity, hence leading to greater chances of value collisions.

In [33], we theoretically determined and experimentally estimated the algorithmic complexity of trivial/simple (denoted here by S) and random ER graphs. Regular graphs, such as completely disconnected or complete graphs, have algorithmic complexity $K(S) = \log |V(S)|$. ER graphs have maximal complexity, so any other complex network is upper-bounded by $K(ER)$ graphs. Now the algorithmic complexity of a BA network is low, because it is based on a recursive procedure, but there is an element of randomness accounted for by the attachment probability.

In [33], theoretical and numerical estimations of algorithmic information content for a range of theoretical and real-world networks are provided. Table 4.1 provides a summary of the theoretical expectations.

To test our numerical approximations to these theoretical estimations we have devised a wide range of experiments, one of the most conclusive with regards to graphs being the one performed on dual and co-spectral graphs. We know that graphs

Table 4.1   Theoretical calculations of K for different network topologies for $0 \le p \le 1$ as derived in [33]. Clearly maximum K is reached for random ER graphs with edge density $p = 0.25$ for which $K(ER) = \binom{|V(ER)|}{2}/2$.

Graph/network	Notation	K
Regular	S	$K(S) = \mathcal{O}(\log |V(S)|)$
Barabási–Albert	BA	$K(BA) = \mathcal{O}(|V(BA)|) + c$
Watts–Strogatz	WS	$\lim_{p \to 0} K(WS) \sim K(S)$
		$\lim_{p \to 1} K(WS) \sim K(ER)$
Random Erdős–Rényi	ER	$K(ER) = \mathcal{O}\left(\frac{n(n-1)}{16p|p-1|}\right)$

and their duals must have about the same algorithmic complexity because there is a computer program of (small) fixed size that can transform any graph into its dual by simple replacement of edges for nodes and nodes for edges. While for duals it was clear that if our methods were numerically sound they would approximate each other, co-spectrality is a less trivial property of graphs. For both tests our methods outperformed other approaches, such as compression and Shannon entropy [35].

4.4   Conclusions

We have surveyed concepts and methods for characterizing graphs and networks using classical and algorithmic information theory, including aspects of lossless compression and Shannon entropy, in particular the capabilities and limitations of the several approaches serviceable at different scales and for different purposes.

One obvious direction of research that we have taken is the investigation of dynamic changes in the information content of networks over time – beyond the study of static networks. We have called this area *algorithmic information dynamics* and it was introduced in [40] based on measures of algorithmic complexity, which was possible thanks to their robust capabilities in characterizing properties of graphs and networks.

References

[1] Zenil H, Badillo L, Hernández-Orozco S, Hernandez-Quiroz F. Coding-theorem like behaviour and emergence of the universal distribution from resource-bounded algorithmic probability. *Int J Parallel Emergent Distrib Syst.* 2018. doi:10.1080/17445760.2018.1448932

[2] Ziv J, Lempel A, A universal algorithm for sequential data compression. *IEEE Trans Inf Theor.* 1977;23(3):337–343.

[3] Zenil H, Small data matters, correlation versus causation and algorithmic data analytics. In J. Wernecke et al.(eds), *Predictability in the World: Philosophy and Science in the Complex World of Big Data*. Berlin: Springer Verlag, forthcoming.

[4] Babai L, Luks E. Canonical labelling of graphs. In: *Proc. 15th ACM Symposium on Theory of Computing*, 1983, pp. 171–183.

[5] Erdös P, Rényi A, On random graphs I. *Publ. Math.* 1959: 6:290–297.

[6] Gilbert EN, Random graphs. *Ann Math Stat.* 1959;30:1141–1144. doi:10.1214/aoms/1177706098.

[7] Zenil H, Soler-Toscano F, Dingle K, Louis A. Graph automorphisms and topological characterization of complex networks by algorithmic information content. *Physica A.* 2014; 404:341–358.

[8] Boccaletti S, et al. The structure and dynamics of multilayer networks. *Phys Rep.* 2014;544(1):1–122.

[9] Chen Z, Dehmer M, Emmert-Streib F, Shi Y, Entropy bounds for dendrimers. *Appl Math Comput.* 2014;242:462–472.

[10] Orsini C, Dankulov MM, Colomer-de-Simón P, et al. Quantifying randomness in real networks. *Nat Commun.* 2015;6:8627.

[11] Bianconi G. The entropy of randomized network ensembles. *Europhys Lett.* 2007;81(2):28005.

[12] Estrada E, José A, Hatano N. Walk entropies in graphs. *Linear Algebra Appl.* 2014;443:235–244.

[13] Dehmer M, Mowshowitz A. A history of graph entropy measures. *Inf Sci.* 2011;181(1):57–78.

[14] Sengupta DC, Sengupta JD. Application of graph entropy in CRISPR and repeats detection in DNA sequences. *Comput Mol Biosci.* 2016;6(3):41.

[15] Korner J, Marton K. Random access communication and graph entropy. *IEEE Trans Inf Theor.* 1988;34(2):312–314.

[16] Dehmer M, Borgert S, Emmert-Streib F. Entropy bounds for hierarchical molecular networks. *PLoS One.* 2008;3(8):e3079.

[17] Dehmer M, Barbarini N, Varmuza K, Graber A, A large scale analysis of information-theoretic network complexity measures using chemical structures. *PLoS One.* 2009;4(12): e8057.

[18] Zenil H, Kiani NA, Tegnér J, Low algorithmic complexity entropy-deceiving graphs, *Phys Rev E.* 2017;96:012308.

[19] Zenil H, Kiani NA, Tegnér J. An algorithmic refinement of Maxent induces a thermodynamic-like behaviour in the reprogrammability of generative mechanisms. In Wolpert D, Kempes C, Grochow J, Stadler P (eds), *The Interplay of Thermodynamics and Computation in Both Natural and Artificial Systems.* Santa Fe, CA: Santa Fe Institute (forthcoming).

[20] Zenil H, Kiani NA, Tegnér J. Algorithmic complexity of motifs, clusters, superfamilies of networks. In: *Proc. IEEE International Conference on Bioinformatics and Biomedicine,* 2013.

[21] Zenil H, Kiani NA, Tegnér J, Quantifying loss of information in network-based dimensionality reduction techniques. *J Complex Netw,* 2016. doi:10.1093/comnet/cnv025.

[22] Chaitin GJ. On the length of programs for computing finite binary sequences. *J ACM.* 1966;13(4):547–569.

[23] Kolmogorov AN. Three approaches to the quantitative definition of information. *Prob Inf Transmission.* 1965;1(1):1–7.

[24] Calude CS. *Information and Randomness: An Algorithmic Perspective.* 2nd edition. New York: Springer, 2010.

[25] Li M, Vitányi P. *An Introduction to Kolmogorov Complexity and Its Applications.* 3rd edition. New York: Springer, 2009.

[26] Turing AM. On computable numbers, with an application to the Entscheidungsproblem. *Proc Lond Math Soc.* 1937;2(42):230–265.

[27] Kirchherr WW, Li M, Vitányi PMB. The miraculous universal distribution. *J Math Intelligencer.* 1997;19:7–15.

[28] Levin LA. Laws of information conservation (non-growth) and aspects of the foundation of probability theory, *Prob Inf Transmission.* 1974;10(3):206–210.

[29] Solomonoff RJ. A formal theory of inductive inference: parts 1 and 2. *Inf Contr.* 1964;7:1–22, 224–254.

[30] Cover TM, Thomas JA. *Elements of Information Theory,* 2nd edition. New York: Wiley-Blackwell, 2009.

[31] Delahaye J-P, Zenil H. Numerical evaluation of the complexity of short strings: a glance into the innermost structure of algorithmic randomness. *Appl Math Comput.* 2012;219:63–77.

[32] Soler-Toscano F, Zenil H, Delahaye J-P, Gauvrit N. Calculating Kolmogorov complexity from the frequency output distributions of small Turing machines. *PLoS One.* 2014;9(5):e96223.

[33] Zenil H, Kiani NA, Tegnér J, Methods of information theory and algorithmic complexity for network biology. *Sem Cell Dev Biol.* 2016; 51:32–43.

[34] Zenil H, Soler-Toscano F, Delahaye J-P, Gauvrit N, Two-dimensional Kolmogorov complexity and validation of the coding theorem method by compressibility. *arXiv:1212.6745 [cs.CC],* 2013.

[35] Zenil H, Soler-Toscano F, Kiani NA, Hernández-Orozco S, Rueda-Toicen A. A decomposition method for global evaluation of Shannon entropy and local estimations of algorithmic complexity. *arXiv:1609.00110[cs.IT],* 2016.

[36] Buhrman H, Li M, Tromp J, Vitányi P. kolmogorov random graphs and the incompressibility method, *SIAM J Comput.* 1999;29(2):590–599.

[37] Langton CG. Studying artificial life with cellular automata. *Physica D.* 1986;22(1–3):120–149.

[38] Albert R, Barabási A-L. Statistical mechanics of complex networks. *Rev Modern Phys.* 2002;74:47.

[39] Milo R, Shen-Orr S, Itzkovitz S, et al. Network motifs: simple building blocks of complex networks. *Science.* 2002;298(5594):824-827.

[40] Zenil H, Kiani NA, Marabita F, et al. An algorithmic information calculus for causal discovery and reprogramming systems. *arXiv:1709.05429,* 2018.

5 Integration and Feature Identification in Multi-layer Networks using a Heat Diffusion Approach

Gordon Ball and Jesper Tegnér

5.1 Introduction

The increased amount and diversity of molecular data offers new opportunities for clinical research and practice, including to identify predictive biomarkers to personalize therapies and to obtain an increased precision in diagnostic procedures. Such features could also be used for an in-depth analysis of the biological processes initiating and driving disease progression. These two challenges have received large and increasing efforts since the sequencing of the human genome. It has become clear that, due to the complexity of the genome and the myriad associated processes, it is not sufficient to enumerate a list of single-nucleotide polymorphisms (SNPs) associated with a particular disease. The predicted amount of risk for disease originating from SNPs is small (below 15%) and offers limited insight into mechanisms of disease. Naturally, there is a need to assess several data types and their integration in order to improve predictive capacity and mechanistic understanding [1–8]. Significant progress has been made in the areas of cancer, cardiovascular and metabolic diseases [9–13].

In part because here we have access to rich data from humans and animal models, as well as the opportunity to perform targeted validation experiments. For example, prediction of key molecular processes for obesity can readily be interrogated using animal models. In contrast, for neurodegenerative diseases, such as Alzheimer's disease or Parkinson's disease, the amount of data is diverse and more limited, and importantly it is more challenging to perform experimental interventions in humans, and the animal models are only marginally relevant for these diseases [14]. Hence, new approaches are needed to identify putative mechanisms and biomarkers of neurodegenerative diseases, to a large extent driven from publicly available data. There are several candidate processes present in publicly available data [15–17]. Here, we use Alzheimer's disease as an example; known mechanisms include genetic variants, early dysregulation events such as inflammation and prions, leading to dementia.

There is a need to prioritize between available candidates, to discover new processes, and finally their putative internal relationships. How to integrate multiple types of data in order to identify 'interesting' entities or features is a classic problem. This challenge is a special case of the general problem of identifying relevant features from a large data set predicting one or several target variables [18]. In the case of multiple data types, this is in practice even more challenging to achieve in an unbiased manner [2, 19].

A common bioinformatics approach to this challenge is to iterate through different types of data, performing filtering at each step for each data type individually, which can unfortunately lead to promising candidates being disregarded early [2, 20]. It is essentially a univariate statistical approach, thus not being amenable to detecting combinations of features – that is, multi-variate characteristics in the data. Ideally one would therefore like to consider all informative sources of data concurrently and choose interesting candidates in a single step. Recently, an approach referred to as a heat flow analysis has been successfully applied in the analysis of large data sets on mutations in cancer [21, 22]. The analysis facilitated the prioritization between the myriad different mutations in pathways.

We therefore reasoned that it could be a promising approach to use for large and different data sets supporting the prioritization of molecular components, using Alzheimer's disease as a case study. The method we explore in the current chapter assumes that the underlying data can be represented as a network, with nodes representing biological entities and edges associations between them. An example would be a network with genes and SNPs as the nodes, and edges representing the SNPs localized within a gene, and measured gene–gene interactions. The network can be either directed or undirected, with the latter case being represented as pairs of equivalent directed edges. We therefore developed a multi-partite graph representation of several different data types (Figure 5.1)

Our integrative network representation constitutes an analogue to physics in that we consider the nodes to be metal spheres connected (edges) by thin metal rods. Heat diffusion is a well studied physical process by which heat flows between bodies at different starting temperatures. The microscopic description is of little relevance here, but if we consider a simplified physical model of two large objects of equivalent mass at temperatures t_1, t_2 connected by a thin connector of much smaller mass and thermal conductivity α, then heat flows from the hotter to the colder body with rate $\alpha \Delta t$. As time passes, heat flows through the link and the two bodies approach temperature equilibrium.

It is easy to imagine in such a scenario that, were one sphere to be heated, the heat would gradually spread out through the network. The idea is then to identify interesting 'hot spots' in the system, corresponding to interesting features, combining several data types. To identify the nodes or subnetworks of interest, heat flow is simulated in the network. A prior distribution of heat is first chosen (such as SNPs already implicated in a disease). The heating is then removed and the network left until the heat distribution reaches a steady state. Hot spots or subnetworks can then be identified, and their significance tested by permuting the starting heat sources or edge weights. Yet, it can readily be observed that it is still necessary to make informed choices in the selection of initial conditions as well as the construction of the network. Here we

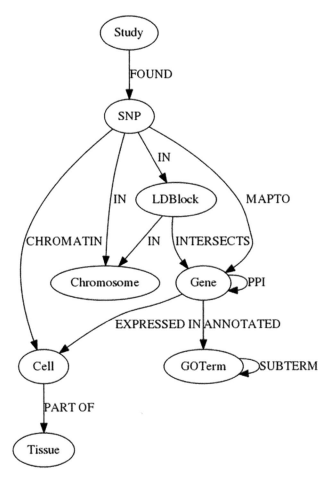

Figure 5.1   Summarized multi-partite graph representation of several different data types in Alzheimer's disease.

construct a network consisting of public data for Alzheimer's disease [14, 16] using the following data types: SNPs, LD blocks, genes, gene ontology (GO) annotations, cell types and tissues. We use SNPs that have previously been implicated as having disease risk, disease-associated GO terms and brain tissues as prior distributions. Running the simulation corresponds to allowing the signals to spread (diffuse) in the network, thus to capture hot spots in the network.

5.2   Mathematical Formulation and Computational Implementation

The network is represented by the system of ordinary differential equations $\frac{dh(t)}{dt} = \alpha W h(t)$, where $h(t)$ is a vector of heat values for each node, α is a global scaling factor, t is the simulation time and W is a matrix containing the edge weights and the negative sum of edge weights on the diagonal. This has an exact solution $h(t) = e^{\alpha t W h(0)}$, but in practice there do not exist good methods to calculate the matrix exponential for non-trivial sized networks, and hence an iterative approach is needed.

The underlying system of interest must be representable as a single network. This can have multiple classes of both nodes and edges. From the network construction, the following information must be extracted:

- The network adjacency matrix, identifying which entities are connected. Edges can be either undirected or directed, with the latter case represented by a pair of equivalent directed edges.
- A conductivity value for each edge. This is a weighting value that needs to be generated somehow based on the underlying properties of the system, with edges that are considered more important or more informative to the question at hand having higher weighting, and edges that are more uncertain or less relevant having lower weighting. How to generate the actual weight values is subject-dependent. Edges with no interesting properties can receive a default weight of 1, but this will reduce the usefulness of the model.
- A set of nodes that will be heat sources. The model starts at time zero with most nodes having zero heat, and a small set being chosen as heat sources. These should be entities already known to be relevant to the question at hand. The magnitude of individual heat sources can be varied according to subject-dependent rules.

Having selected the network structure, weights for the edges and starting sources of heat, the simulation is executed. Heat will diffuse away from the original sources and spread across the network. Early in the simulation the starting sources will remain the hottest nodes, along with those near them; mid-way through the simulation, nodes with high network centrality and high edge conductivities are likely to dominate, and finally the heat distribution will approach equilibrium. At an appropriate point, the simulation is terminated and the ranking of heat values read out.

The appropriate stopping time will depend on the priority that is wished to be placed on the different contributing factors; shorter integration times will favour proximity to heat sources and later integration times will favour network structure and edge weighting as key factors. The actual values of the time parameter will depend on the distribution of conductivity values used, and the effective diameter of the network.

5.3   Results

5.3.1   Construction of an Alzheimer's Network

We applied this method by constructing a model of Alzheimer's disease. Our objective was to identify in an unbiased way new and potentially interesting biomarkers. The choice to work with this type of model was made precisely because a previous, conventional filtering approach was criticized for making unclearly motivated choices and focusing on narrow regions of possibilities.

We identified several relevant types of public data: SNPs found by genome-wide association studies; genes from which biomarkers could be derived; linkage-disequilibrium blocks grouping genetic entities; functional and pathway annotations of genes; and different cell and tissue types. These were further connected by other types of data; protein–protein interactions to link genes; genomic position linking SNPs, genes and LD blocks; gene expression data connecting genes and cell types; and chromatin state data connecting SNPs and cell types.

These combinations of data yielded a network containing approximately 30,000 entities and 200,000 edges. For each type of association we chose rules to generate a weight; for example, SNP-to-gene edges were weighted with a fixed value if the SNP was intronic, and a linear falloff if the SNP was within 15 kb of either the 3′ or 5′ ends of the gene region; gene-to-cell edges were weighted according to the log2 difference from the median expression for the cell type, and SNP-to-cell edges were assigned fixed weights according to the chromatin state, with higher values given for strong enhancer or promoter states.

We chose as our starting heat sources SNPs that had been implicated by genome-wide association studies (weighted by odds ratio, where available), a manually curated list of functional terms associated with the major disease processes, and the set of body tissues known to be significant for the diseases in question.

The network was stored and queried using the neo4j graph database (www.neo4j .com), which allows us to query the network in a topological manner. The actual heat diffusion model was written in Python, using the numpy and scipy libraries for efficient arrays and numerical integration, respectively. The kernel function was subsequently rewritten using the Cython dialect of Python, which could be efficiently compiled to C, achieving at least a factor 10 speed increase, and the memory footprint was reduced from $\sim$8 GB for a 30 000 × 30 000 array to <1 GB using a sparse representation.

Appropriate integration time was selected by computing histograms of the heat distribution as the simulation ran and choosing a point at which the initial bimodal distribution (a small number of hot nodes and all remaining nodes cold) appeared to merge into a single population, which should combine contributions from starting heat sources and the network topology.

5.3.2 Simulation of the Alzheimer's Model

Running the model generated, one the one hand, a grand overview of the whole system in which some candidates can be used as anchor points for visualization (Figure 5.2).

On the other hand, the results can readily be inspected as a list of genes ranked by the heat in the graph after the simulation has been stopped. Such a list contains a number of expected results (genes with known disease associations), as shown in Figure 5.3.

However, we encountered a persistent issue that well-known and highly significant genes never appeared high in our ranked list. This appears to largely have been a consequence of the process of generating edge weights by reduction of other data; while we could attempt to represent particularly significant features

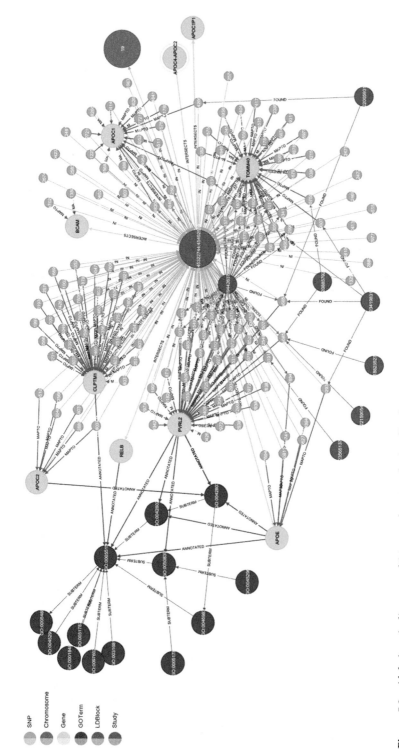

Figure 5.2 Alzheimer's disease multi-type sub-network visualization.

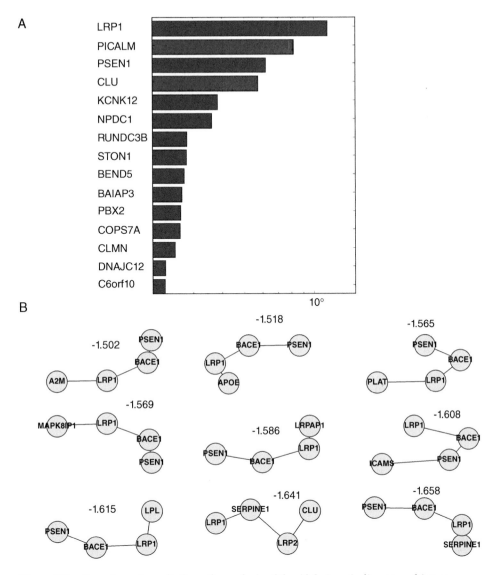

Figure 5.3   Top relevant results from the analysis of the Alzheimer's disease multi-type network. (a) Top ranked genes. (b) Top-ranked hot spots.

with large weights or starting heat values, the presence of a small number of highly weighted or heated features in a very large network would quickly average out and the distinctiveness of the feature would be lost.

5.3.3   Estimation and Assessment of the Influence of Different Simulation Parameters

Next, we evaluated the influence of the different factors (network structure, edge weighting, heat sources) on the final result (i.e. list of genes) by performing permutation experiments and observing the degree of change in the resulting rankings.

By re-running the simulation, having either randomly swapped edges in the adjacency matrix, randomly swapped edge weights or randomly swapped the locations of starting heat sources, we can get an idea of the relative importance of these factors to the final rankings. The specific effects of particular classes of edge or node can be tested by re-running the network with them completely removed (or at least disconnected), and the effect of individual, highly network-centric nodes tested similarly. If the network and heat overlay have been well implemented, all of these modification should cause a detectable difference but none should cause an overwhelming one.

Furthermore, there are a large number of implicit parameters involved in the heat diffusion model, where the major one is actually inherent to the heat diffusion process (the integration time). The remaining parameters, as discussed above, arise from the methods used to assign edge weights and heat source magnitudes, both whichever coefficients are used in the formulae for generating those values and the overall scaling factors for different classes of edges and heat sources. To assess the utility of a heat flow approach we therefore considered to what extent we could formulate a methodology balanced between overfitting and arbitrary selection of parameters. Thus, in our hands, the ensemble of parameters in the model, listed above, suggests an optimization approach aiming to find a 'better' set of parameters for the specific problem at hand.

This of course introduces an additional point of bias to the model; by trying to optimize the model's output to match some set of expectations, we reduce the value of the method as an 'unbiased' approach to integrating data, but it may be possible to walk a line between overfitting and providing some correction for what may have otherwise been entirely arbitrary parameter choices.

For the Alzheimer's disease model, we set out to improve the number of known biomarkers found near the top of the ranking list in order to improve the credibility of other targets that we might find. This of course is risky; we might only find new targets that are very closely associated with existing ones and hence of limited new interest. To try to avoid the risk of overfitting the model, we adopted a weak scoring criteria in which putative parameter sets would be scored based on how many of a list of 10 known targets appeared in the top 30 results.

The parameter set for this model contained an overall scaling factor for each class of edge, for each class of heat source and the integration time, giving a total of 15 parameters; a relatively simple optimization problem, which was fortunate since each generated parameter set requires the evaluation of the full heat model, with significant CPU and memory requirements.

The optimization was performed using a particle swarm method [23], which provides a good mixture of exploration for new minima and exploitation of found ones, and since it does not attempt to calculate gradients it is relatively insensitive to the non-differentiable nature of our parameter landscape (since the scoring generates only integers). It is also amenable to parallelization, which is appealing in this case where we needed to run a large number of independent but expensive computations.

However, despite expending significant CPU time on optimization, we were not able to find parameter sets producing results significantly better than that produced by the 'naive' parameter set.

5.4 Discussion

Here we have shown how to adapt and use the heat flow method in the context of data integration and prioritization of features. Our analysis produced several candidate genes suggested to be associated with Alzheimer's diseases. Yet, our implementation and analysis revealed several hurdles that need additional consideration in the future.

In brief, the method has several limitations. Potentially rich data about associations between entities has to be reduced to a single scalar value (the edge weight), and a network-based representation may be a too severe simplification over more fuzzy underlying data. It is not obvious how to choose the optimal simulation time after which to obtain results, since such a choice necessarily prioritizes different factors that the model encompasses. The method will perform poorly on networks that are either too sparse (in which case the result will be determined wholly by the connectivity) or too dense (in which case the model will too quickly reach a flat equilibrium). It is not obvious if there exists a sweet spot between these two extremes.

Moreover, as we have to make a number of choices in constructing the model, such as how to weight the edges, the position and magnitude of starting heat sources and the simulation time to elapse before reading our results, the method cannot be said to be wholly unbiased, but the advantage compared to more traditional iterative methods should be that everything is at least considered concurrently, and while choices have been made, we have to clearly state the choices that have been made.

In the case of our neurodegenerative disease model, we were ultimately forced to conclude that the use of this model provided little additional insight into the problem. The results, both from the naive model and the best solutions found by optimization, lacked credibility due to their failure to identify several well-known risk factors.

There are several possible reasons for this:

- Reliable, subject-specific knowledge cannot be represented sufficiently strongly in the model. In this case, the significance of a small number of extremely high-risk SNPs could not adequately be injected into the network, since even if they were assigned extremely large heat values, the effect still diffused too quickly into the general network and the knowledge did not sufficiently contribute to the result.
- In this case, the network was wholly undirected. Even directed subsets of the network would produce more meaningful heat flow patterns during the diffusion phase rather than just a symmetrical spread, which tends more quickly to flat equilibrium. Unfortunately, it is not obvious how to translate many of the underlying undirected relations to a directed edge.
- Different edge classes with significantly different numbers of edges are likely to dominate, even if significant weight scaling is applied. In this model, SNP-to-cell and cell-to-gene edges were responsible for the majority of the network edges, and even with reduced weights, tended to dominate the, for example, SNP-to-gene bandwidth despite not being direct.
- The selection of data types incorporated into the underlying network may just not have ultimately been suitable for the inference problem at hand, and the choice of method might not have been the principal weakness.

- The method is linearly separable with respect to each individual heat source, since there are no saturation or other non-linear effects involved. Thus, we are not so much achieving an integration of the different heat sources used, but rather effectively just summing their contributions. This is important if the problem at hand specifically addresses synergies between the different contributions used as heat sources. It would be possible to extend the model with saturation effects on either edges or nodes to mitigate this, but this would introduce further parameters to the model.
- Particularly when using parameter sets that might induce order-of-magnitudes difference in edge weights or heat sources, numeric stability can be an issue. This is generally solvable by configuration of the integration routine used, but the trade-off for higher accuracy tends to be much longer runtime, which is particularly likely to be prohibitive if including an optimization step.

The approach discussed above originates from the standpoint of using the heat network to generate hypotheses; it is also possible to invert the problem and rather than ranking entities in the network (implicitly treating every node as a hypothesis), one may instead take a list of hypotheses generated by some other method and use the heat network to rank them according to the heat found by mapping them into the heat network. This approach may be more useful than a-priori hypothesis generation since it avoids the possibility of false positives generated by the heat network, and still is able to distil some of the useful properties of the network into a ranking of the supplied hypotheses. This is an option that could be explored in future studies.

The method is not generally invalid; indeed, it has been used for some successful biological problems, albeit they have not involved integration between different data types; however, the considerations above must be considered when using it for data integration problems.

References

[1] Tegner JN, Compte A, Auffray C, et al. Computational disease modeling: fact or fiction? *BMC Syst Biol.* 2009;3:56.

[2] Gomez-Cabrero D, Abugessaisa I, Maier D, et al. Data integration in the era of omics: current and future challenges. *BMC Syst Biol.* 2014;8(Suppl 2):I1.

[3] Coveney PV, Diaz-Zuccarini V, Graf N, et al. Integrative approaches to computational biomedicine. *Interf Focus.* 2013;3(2):20130003.

[4] Tan K, Tegner J, Ravasi T. Integrated approaches to uncovering transcription regulatory networks in mammalian cells. *Genomics.* 2008;91(3):219–231.

[5] Cano I, Lluch-Ariet M, Gomez-Cabrero D, et al. Biomedical research in a digital health framework. *J Transl Med.* 2014;12(Suppl 2):S10.

[6] Ahmad A, Fröhlich H. Integrating heterogeneous omics data via statistical inference and learning techniques. *Genomics Comput Biol.*2016;2(1).

[7] Roca J, Cano I, Gomez-Cabrero D, Tegnér J. From systems understanding to personalized medicine: lessons and recommendations based on a

multidisciplinary and translational analysis of COPD. *Meth Mol Biol.* 2016;1386:283–303.

[8] Richardson S, Tseng GC, Sun W. Statistical methods in integrative genomics. *Annu Rev Stat Appl.* 2016;3:181–209.

[9] Sanchez-Garcia F, Villagrasa P, Matsui J, et al. Integration of genomic data enables selective discovery of breast cancer drivers. *Cell* 2014;159:1461–1475.

[10] Schadt EE. Molecular networks as sensors and drivers of common human diseases. *Nature.* 2009;461(7261):218–223.

[11] Neto EC, Broman AT, Keller MP, et al. Modeling causality for pairs of phenotypes in system genetics. *Genetics.* 2013;193:1003–1013.

[12] Hägg S, Skogsberg J, Lundström J, et al. Multi-organ expression profiling uncovers a gene module in coronary artery disease involving transendothelial migration of leukocytes and LIM domain binding 2: the Stockholm Atherosclerosis Gene Expression (STAGE) study. *PLoS Genet.* 2009;5(12).

[13] Gomez-Cabrero D, Compte A, Tegner J. Workflow for generating competing hypothesis from models with parameter uncertainty. *Interf Focus.* 2011;1(3):438–449.

[14] Hofmann-Apitius M, Ball G, Gebel S, et al. Bioinformatics mining and modeling methods for the identification of disease mechanisms in neurodegenerative disorders. *Int J Mol Sci.* 2015;16(12).

[15] Bateman RJ, Xiong C, Benzinger TLS, et al. Clinical and biomarker changes in dominantly inherited Alzheimer's disease. *N Engl J Med.* 2012;367(9):795–804.

[16] Kumar A, Singh A. A review on Alzheimer's disease pathophysiology and its management: an update. *Pharmacol Rep Instit Pharmacol, Polish Acad Sci.* 2015;67(2):195–203.

[17] Huang Y, Mucke L. Alzheimer mechanisms and therapeutic strategies. *Cell.* 2012;148(6):1204–1222.

[18] Nilsson R, Peña JM, Björkegren J, Tegnér J. Consistent feature selection for pattern recognition in polynomial time. *J Mach Learn Res.* 2007;8.

[19] Ma'ayan A, Rouillard AD, Clark NR, et al. Lean big data integration in systems biology and systems pharmacology. *Trends Pharmacol Sci.* 2014;35(9):450–460.

[20] Gomez-Cabrero D, Tegner J. Data integration: towards understanding biological complexity. In: Stumpf M, Balding DJ, Girolami M (eds), *Handbook of Statistical Systems Biology.* Chichester: Wiley, 2011.

[21] Vandin F, Upfal E, Raphael BJ. Algorithms for detecting significantly mutated pathways in cancer. *J Comput Biol.* 2011;18(3):507–522.

[22] Leiserson MDM, Vandin F, Wu H-T, et al. Pan-cancer network analysis identifies combinations of rare somatic mutations across pathways and protein complexes. *Nat Genet.* 2015;47(2).

[23] Bonyadi MR. Particle swarm optimization for single objective continuous space problems: a review. *Evol Comput.* 2017;25(1):1–54.

PART III
MULTI-LAYER NETWORKS

6 Multiplex Networks

Ginestra Bianconi

6.1 Introduction

Decoding the entire human genome has been undoubtedly a pivotal scientific achievement at the turn of this century. However, the great expectations beyond the release of the first complete map of the genome had to face the complexity of the biological logic that goes beyond the "one gene–one disease" schematic picture that was dominating the scientific debate at that time. Now we know that most of the major diseases are complex and that the map between the phenotype and the genotype is highly affected by the network of networks, including all molecular interactions in the cell. It is therefore essential to adopt a network science approach and to characterize the complex web of all molecular interactions in the cell.

In this direction at the end of the twentieth century the first high-throughput experiments aimed at mapping all the interactions between the constituents of the cell have signaled a turnover in molecular biology. In these last 20 years, thanks to the new availability of high-throughput molecular network data and the network science approach, we have gained unprecedented knowledge of the molecular networks of cells. However, in an order of increasing complexity, initially the single-molecule experiments have been substituted by network science of high-throughput experiments restricted to a single type of biological interaction. In particular, many works have specialized in the topology and dynamics of the protein–protein interaction networks or else the metabolic, transcription or signaling networks.

Only recently has the field of network medicine [1] been established, and new research works [2] have started to combine different types of molecular interactions into a single multi-layer network [3, 4], the interactome, that includes all the interactions of a cell. Therefore, developing and using new methods to extract information from multi-layer molecular networks is key to characterizing and identifying the properties of the interactome connected to major diseases.

The multi-layer network perspective is not only essential for the further development of network medicine but it also play a crucial role where one wants to combine and compare different molecular cell biology experiments [5–7] to gain new relevant information about the biological problem under study.

Despite the fact that in network science the study of molecular networks using multi-layer network methods is only in its infancy, in the context of brain networks [8, 9] multi-layer networks have already been used extensively to characterize the different interactions between neurons (such as synaptic connection or gap junctions) [10–12], and at a more macroscopic scale they have been used to characterize the functional brain network between brain regions [13]. Therefore, these studies constitute a proof of concept of the possible advantage of adopting a multi-layer network perspective for the study of biological networks.

In this chapter we will review the major properties of multiplex networks. Multiplex networks are the simplest examples of multi-layer networks. They are networks formed by a given set of N nodes having interactions of different types. The interactions of the same type form a network called a *layer* of the multiplex network. Therefore, a multiplex networks of M layers is formed by interactions of M different connotations.

Biological real multiplex networks can describe coupled protein–protein interaction networks and gene–gene correlation networks where one makes the simplification that one gene corresponds to one protein. Additionally, multiplex networks can be constructed by combining different gene–gene correlations networks extracted from gene expression in different conditions (see, for example, the multiplex networks used in [5–7]). Finally, the characterization of multiplex networks opens new perspectives for the study of the full interactome network [2]. We note, however, that it is also possible to go beyond the multiplex network framework and extract very useful information about cancer data from multi-layer networks formed by a gene–gene network and protein interaction networks where the links across layers are not exclusively one-to-one [14].

This chapter aims at providing the major properties of multiplex networks and the major tools used to extract relevant information from their structure. Examples from the study of molecular networks and brain networks will illustrate the relevance of the illustrated multiplex network analysis.

6.2   Definition and Basic Properties

A multiplex network with $\vec{G}$ given by

$$\vec{G} = (G^{[1]}, G^{[2]}, \ldots G^{[M]}) \tag{6.1}$$

is formed by M layers and N nodes [3, 4, 15] . Each layer $\alpha = 1, 2, \ldots, M$ consists of a network

$$G^{[\alpha]} = (V, E^{[\alpha]}) \tag{6.2}$$

formed by a set of nodes V and a set of links $E^{[\alpha]}$. The set V of N nodes $i = 1, 2, \ldots, N$ is identical for all layers. The set of links $E^{[\alpha]}$ is instead specific for each layer α.

Examples of multiplex networks include, for instance, a multiplex network of $M = 3$ layers and N nodes (genes) constructed by combining information coming

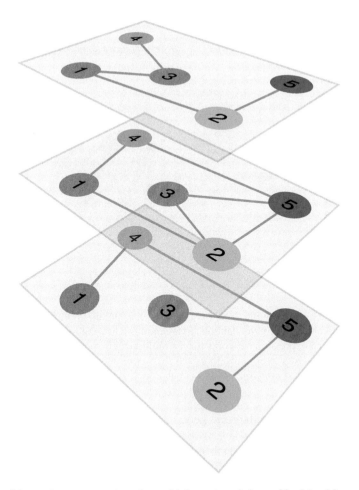

Figure 6.1 Schematic representation of a multiplex network formed by $M = 3$ layers. These three layers could, for instance, represent gene–gene expression, transcription factor co-regulation and micro-RNA co-regulation as in [5]. Reprinted figure from [16] under the Creative Commons Attribution 4.0 International License.

from gene–gene expression, transcription factor co-regulation and micro-RNA co-regulation as in [5] (see the schematic representation in Figure 6.1) or the multiplex network with many layers indicating gene–gene co-expression networks in different tissues studied in [6] (see Figure 6.2). Finally, multiplex networks can describe temporal networks when the different layers indicate interactions measured in different time windows such as the functional brain networks studied in [13].

Sometimes it is useful to distinguish between node i in layer α indicated as (i, α) and node i in layer α' indicated as (i, α'). The nodes (i, α) with $\alpha = 1, 2, \ldots, M$ are called *replica nodes*.

Usually, replica nodes indicate the same identity, for instance they might indicate the same gene in different contexts. However, there are cases in which replica nodes can indicate different identities, for instance genes in one layer and proteins in another

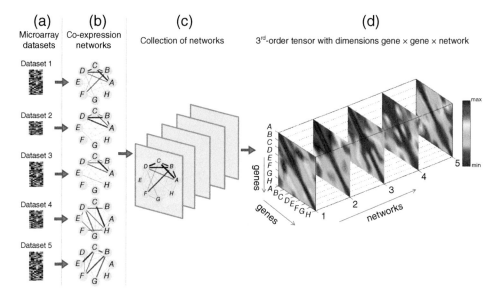

Figure 6.2   Illustration of the multi-layer (tensorial) representation for multiple co-expression networks. Microarray datasets (a) are modeled as a collection of co-expression networks (b); these co-expression networks form a multi-layer network (c) that can be represented as a third-order tensor such that each slice represents the adjacency matrix of one network (d). Reprinted from [6]. © 2011 Li et al.

layer, as long as we assume that one gene of the first layer corresponds to a single protein in the second layer.

It is sometimes useful to connect replica nodes by links called *interlinks* and to attribute to these interlinks a different role/dynamics. However, interlinks might be sometimes considered somewhat artificial and not really encoded on biological multiplex network datasets. Therefore, since a multiplex network is determined uniquely by its layer composition and the one-to-one mapping between replica nodes, here we will focus on multiplex network properties that are induced by this fundamental structure; that is, in absence of interlinks. We refer the reader interested in the analysis of multi-layer networks with interlinks to the chapters that follow.

Here and in the following we will indicate with $\mathbf{a}^{[\alpha]}$ the adjacency matrix of each layer $\alpha = 1, 2 \ldots, M$ that can be eventually weighted and/or directed. However, for simplicity in the following we will consider only undirected and unweighted multiplex networks unless stated otherwise. In this case the matrix elements are given by

$$a_{ij}^{[\alpha]} = \begin{cases} 1 & \text{if node } i \text{ is connected to node } j \text{ in layer } \alpha, \\ 0 & \text{otherwise.} \end{cases} \tag{6.3}$$

Alternatively a multiplex network can be represented by a tensor $\mathcal{T}$ of elements

$$\mathcal{T}_{ij\alpha} = a_{ij}^{[\alpha]}. \tag{6.4}$$

For multiplex networks, even basic network measures acquire different properties. For instance the same definition of degree is modified and one can assign to each node i a *multiplex degree* $\mathbf{k}_i$ given by

$$\mathbf{k}_i = \left(k_i^{[1]}, k_i^{[2]}, \ldots, k_i^{[M]} \right), \tag{6.5}$$

where $k_i^{[\alpha]}$ is the degree of node i in layer α. Therefore the multiplex degree is no longer a scalar (a single number) like in single networks, but is a vector whose components are the degree of the node on each separate layer.

The information that can be extracted from a multiplex network is not simply reducible to the information that is carried by its single layers taken in isolation. In fact, a multiplex network includes many built-in correlations that are a resource for multiplex network analysis and multiplex network inference. One major aspect of multiplex network data is the tendency of some layers to enhance (or suppress) the number of interactions that exist between the same set of nodes in different layers. In the presence of a significant number of these interactions we say that in the multiplex network there is a significant *link overlap*.

To quantify the significance of link overlap it is possible to associate to each pair of layers of a multiplex networks the *total overlap* $O^{[\alpha,\alpha']}$ given by the number of pairs of nodes connected in both layers, that is

$$O^{[\alpha,\alpha']} = \sum_{i<j} a_{ij}^{[\alpha]} a_{ij}^{[\alpha']}. \tag{6.6}$$

Additionally, one can define for each node i of the multiplex network the *local overlap* $o_i^{[\alpha,\alpha']}$ providing a local measure of overlap and indicating the number of nodes connected to node i in both layers, that is

$$o_i^{[\alpha,\alpha']} = \sum_{j=1}^{N} a_{ij}^{[\alpha]} a_{ij}^{[\alpha']}. \tag{6.7}$$

These quantities are sometimes normalized, giving rise to the following definition of the normalized total overlap $\hat{O}^{[\alpha,\alpha']}$ and the normalized local overlap $\hat{o}_i^{[\alpha,\alpha']}$ given by

$$\hat{O}^{[\alpha,\alpha']} = \frac{\sum_{i<j} a_{ij}^{[\alpha]} a_{ij}^{[\alpha']}}{\sum_{i<j} \left(a_{ij}^{[\alpha]} + a_{ij}^{[\alpha']} - a_{ij}^{[\alpha]} a_{ij}^{[\alpha']} \right)},$$

$$\hat{o}_i^{[\alpha,\alpha']} = \frac{\sum_{j=1}^{N} a_{ij}^{[\alpha]} a_{ij}^{[\alpha']}}{\sum_{j=1}^{N} \left(a_{ij}^{[\alpha]} + a_{ij}^{[\alpha']} - a_{ij}^{[\alpha]} a_{ij}^{[\alpha']} \right)}. \tag{6.8}$$

From these normalized measures it is possible to assess whether or not two layers display significant overlap with respect to a uniform hypothesis. However, we will discuss a more statistically relevant way to assess the significance of link overlap with respect to a null hypothesis when discussing multilinks in Section 6.3.

Very often a multiplex network analysis compares intrinsically proper multiplex network measures such as total and local overlap to corresponding properties of a single so-called *aggregated network*. The aggregated network does not distinguish between the different nature of the interactions of different layers and provides a single network reference point for the description of the multiplex network. In particular, the aggregated network of a multiplex network is a single network

$$\tilde{G} = (V, \tilde{E}) \tag{6.9}$$

where any pair of nodes is connected if they are connected in at least one layer of the multiplex network. Therefore the adjacency matrix $\tilde{a}$ has elements given by

$$\tilde{a}_{ij} = \begin{cases} 1 & \text{if} \quad \sum_{\alpha=1}^{M} a_{ij}^{[\alpha]} > 0, \\ 0 & \text{otherwise.} \end{cases} \tag{6.10}$$

Even if the original multiplex network is unweighted, the links (i, j) of the aggregated network can be assigned to a natural weight given by the *link multiplicity* defined as the number of layers in which node i is connected to node j, that is

$$v_{ij} = \sum_{\alpha=1}^{M} a_{ij}^{[\alpha]}. \tag{6.11}$$

The strength S_i of a node i in the aggregated network is given by the sum of the weights of its links, that is

$$S_i = \sum_{j=1}^{N} v_{ij}, \tag{6.12}$$

and its degree K_i is instead given by the number of its links in the aggregated network, that is

$$K_i = \sum_{j=1}^{N} \tilde{a}_{ij}. \tag{6.13}$$

Note that due to the fact that two nodes can be connected in multiple layers, the sum of the degree of a node in every layer of the multiplex network is equal to its strength in the aggregated network, not as one could naively assume to its degree. In fact, we have

$$S_i = \sum_{j=1}^{N} v_{ij} = \sum_{j=1}^{N} \sum_{\alpha=1}^{M} a_{ij}^{[\alpha]} = \sum_{\alpha=1}^{N} k_i^{[\alpha]}. \tag{6.14}$$

6.3   Multilink and Multidegree

While in single networks each pair of nodes can be either connected or not connected, in multiplex networks two nodes can be connected in multiple ways among the M layers. Therefore it is essential to extend the concept of link to the concept of *multilinks* [15], which fully characterizes in which layers two nodes are connected (see Figure 6.3). A multilink $\vec{m}$ given by

$$\vec{m} = \left(m^{[1]}, m^{[2]}, \ldots, m^{[\alpha]}, \ldots, m^{[M]} \right), \tag{6.15}$$

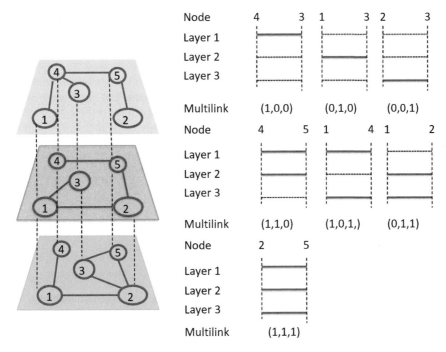

Figure 6.3   Schematic representation of a multiplex network with $M = 3$ layers and its different types of multilinks. Reprinted figure from Ref. [17]. © 2016 American Physical Society.

has elements $m^{[\alpha]} \in \{0,1\}$. Given a pair of nodes (i,j) of an unweighted multiplex network, we say that they are connected by a multilink

$$\vec{m}_{ij} = \left(a_{ij}^{[1]}, a_{ij}^{[2]}, \ldots, a_{ij}^{[\alpha]} \ldots, a_{ij}^{[M]} \right),$$ (6.16)

i.e. if they are connected in each layer α in which $m_{ij}^{[\alpha]} = 1$. Therefore, every pair of nodes is connected by a single multilink and in total there are 2^M possible multilinks. Note that the multilink $\vec{m} = \vec{0}$ indicates the trivial multilink (no connections among the considered pair of nodes) while all the remaining $2^M - 1$ types of multilinks indicate different types of connections. We call the latter *non-trivial multilinks*.

Multilinks can be studied as particularly interesting motifs for multiplex networks by evaluating their significance with respect to suitable null models of multiplex network structures (for a description of null models, see Section 6.5). To this end, one can evaluate their zeta-score with respect to the null hypothesis and extract very important information on the significance of a given type of multilink defining either over-represented multilinks or under-represented ones. An example of such analysis has been conducted in [10] for the multi-layer connectome of C. *elegans*, including synaptic connections, gap junctions and also other monoamide and neuropeptide mediated interactions between neurons extracted from gene expression data (see Figure 6.4). This study reveals significant properties of the multiplex network structure coupled with its dynamical function.

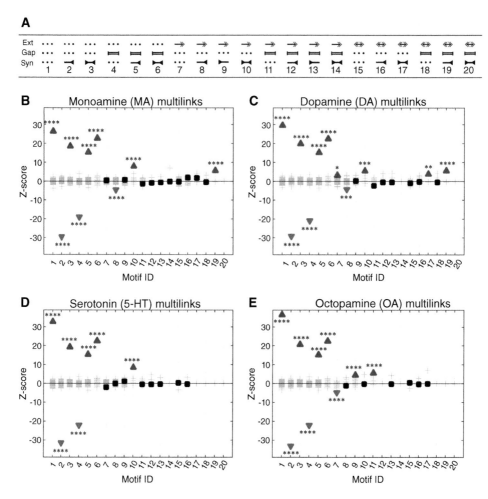

Figure 6.4 The different types of multilinks in the *C. elegans* multi-layer connectome are here used as motifs evaluating their z-score with respect to suitable null models of multiplex networks revealing important features of the multiplex network structure. Reprinted figure from [10] © 2016 Bentley et al.

From multilinks it is possible to define *multidegrees*, generalizing the concept of link overlap to more than two layers. The multidegree $k_i^{\vec{m}}$ of node i is the number of multilinks $\vec{m}$ incident to a node i and can be mathematically expressed as

$$k_i^{\vec{m}} = \sum_{j=1}^{N} \prod_{\alpha=1}^{M} \left[m^{[\alpha]} a_{ij}^{[\alpha]} + (1 - m^{[\alpha]})(1 - a_{ij}^{[\alpha]}) \right]. \tag{6.17}$$

In a duplex network in which one layer is gene–gene co-expression and the other is transcription factor co-regulation, the multidegree $k_i^{(1,1)}$ indicates how many genes are simultaneously co-expressed with node i and regulated by the same transcription factor, $k_i^{(1,0)}$ indicates how many genes that are not co-regulated with gene i are co-expressed to node i, and $k_i^{(0,1)}$ indicates how many (if any) genes are co-regulated with gene i but are not co-expressed. These non-trivial multidegrees can be written as

$$k_i^{(1,0)} = \sum_{j=1}^{N} a_{ij}^{[1]} \left(1 - a_{ij}^{[2]}\right),$$

$$k_i^{(0,1)} = \sum_{j=1}^{N} \left(1 - a_{ij}^{[1]}\right) a_{ij}^{[2]},$$

$$k_i^{(1,1)} = \sum_{j=1}^{N} a_{ij}^{[1]} a_{ij}^{[2]}. \tag{6.18}$$

Interestingly, the multidegree $k_i^{(1,1)}$ in a duplex network of two layers reduces to the definition of the local overlap between the two layers in Eq. (6.7). However, in general the multidegree generalizes the concept of local overlap between two layers to an arbitrary pattern of connections also among more than two layers.

The multidegrees are key characteristics of the nodes of a multiplex network and a relevant centrality measure [18]. In the context of analysis of financial correlations between traded assets, a multiplex network has been constructed by building different layers on the basis of different measures of correlations (linear and nonlinear correlations). In this multiplex network analysis it has been shown that the multidegree is a good centrality measure that can reveal the role of different financial sectors in financial crises. Interestingly, this type of multiplex network analysis can be extended also to the study of biological networks constructed starting from correlations matrices.

The characterization of a multiplex network in terms of multilinks can be used also for compressing the representation of a full unweighted multiplex network in a single weighted matrix. In fact it is possible to consider the matrix $\tilde{a}$ of elements

$$\tilde{a}_{ij} = \sum_{\alpha=1}^{M} 2^{\alpha-1} a_{ij}^{[\alpha]}, \tag{6.19}$$

where each single multilink is indicated by the corresponding binary number with coefficients $m_{ij}^{[\alpha]} = a_{ij}^{[\alpha]}$.

Finally, we observe that in weighted multiplex networks the weights of the links can be affected by the presence of the link overlap [7, 19]. For instance, in a duplex network formed by scientists who collaborate with each other and cite each other, it is possible to detect that the way scientists cite each other is strongly affected by whether they also collaborate with one another or not. These important weight-overlap correlations can be captured by investigating the behavior of the multistrength [19]. The multistrength $s_i^{\vec{m},[\alpha]}$ of node i in layer α is given by the sum of the weights of the links in layer α that belong to a multilink $\vec{m}$, that is

$$s_i^{\vec{m},[\alpha]} = \sum_{j=1}^{N} w_{ij}^{[\alpha]} \delta(\vec{m}, \vec{m}_{ij}), \tag{6.20}$$

where $w_{ij}^{[\alpha]}$ is the weight of the link (i, j) in layer α and $\delta(\vec{m}, \vec{m}_{ij}) = 1$ if $\vec{m} = \vec{m}_{ij}$ and $\delta(\vec{m}, \vec{m}_{ij}) = 0$ otherwise.

The average multistrength $s^{\vec{m},[\alpha]}$ of nodes with given multidegree $k_i^{\vec{m}}$ typically displays the functional behavior

$$s^{\vec{m},[\alpha]} \propto \left(k_i^{\vec{m}}\right)^{\beta^{\vec{m},[\alpha]}}, \tag{6.21}$$

with exponents $\beta^{\vec{m},[\alpha]} \geq 1$. If $\beta^{\vec{m},[\alpha]} = 1$, it means that the weights of the links (belonging to multilinks $\vec{m}$) have a uniform distribution in the network. On the contrary, if $\beta^{\vec{m},[\alpha]} > 1$ it means that nodes with larger multidegree $k_i^{\vec{m}}$ have typically links (belonging to multilinks $\vec{m}$) with larger weight that nodes with smaller multidegree $k_i^{\vec{m}}$.

Whereas the exponents $\beta^{\vec{m},[\alpha]}$ depend significantly on $\vec{m}$, there is a significant correlation between weights and multilinks, signaling a statistically significant way in which the weights of the links are distributed as a function of the presence of different patterns of connections among the nodes.

6.4   Other Multiplex Network Correlations

Link overlap in different layers and weight-overlap correlations are not the only ways information is encoded in a multiplex network structure. Other notable correlations in multiplex networks are inter-layer degree correlations and activities of the nodes [3, 20, 21].

6.4.1   Inter-layer Degree Correlations

Given two layers, the *inter-layer degree correlations* evaluate the tendency of a random node to have high degree in both layers (positive inter-layer degree correlations) or high degree in one layer and low degree in the other (negative inter-layer degree correlations) [20]. The inter-layer degree correlations are an exclusive multiplex network property and differ from the *intra-layer degree corrections* existing between the degrees of linked nodes within each layer. The inter-layer degree correlations can be measured using the average degree in layer α given the degree of the same node in layer α', that is

$$\left\langle k^{[\alpha]}|k^{[\alpha']}\right\rangle = \sum_{k^{[\alpha]}} k^{[\alpha]} P(k^{[\alpha]}|k^{[\alpha']}), \tag{6.22}$$

where $P(k^{[\alpha]}|k^{[\alpha']})$ is the probability that a random node has degree $k^{[\alpha]}$ in layer α given that it has degree $k^{[\alpha']}$ in layer α'. If the function $\left\langle k^{[\alpha]}|k^{[\alpha']}\right\rangle$ is an increasing function of $k^{[\alpha']}$, the inter-layer correlations are positive and hubs of one layer tend to be also hubs of the other; if $\left\langle k^{[\alpha]}|k^{[\alpha']}\right\rangle$ is a decreasing function of $k^{[\alpha']}$, the inter-layer correlations are negative and hubs of one layer tend to be nodes of low degree in the other. However, it is possible that the function $\left\langle k^{[\alpha]}|k^{[\alpha']}\right\rangle$ is not monotonic. In this situation inter-layer correlations can be measured with Pearson, Spearman or Kendall correlation coefficients providing a single number that quantifies the overall correlations among all the nodes of the multiplex network. However, the Pearson correlation coefficient can be dominated by the degree of the high-degree nodes in the presence of high heterogeneities of the degree sequence, the rank Spearman correlation coefficient can be affected by the indetermination of the ranks due to degeneracies of the degree of the nodes. On the contrary, the Kendall correlation coefficient is based on rank and takes into account degree ties; therefore, it is probably the most unbiased global measure of inter-layer degree correlations [3].

We need to note that inter-layer degree correlations can be present simultaneously with intra-layer degree correlations measuring the correlations among the degrees of connected nodes in single layers.

6.4.2   Activities of the Nodes

A multi-layer network is formed by N nodes and M layers. However, not all the nodes might be connected in each layer. In order to capture which nodes are *active* (i.e. connected) in which layer it is possible to characterize a multiplex network using a bipartite network of nodes and layers [22] of incidence matrix $\mathbf{b}$ of elements $b_{i\alpha}$ indicating which node i is connected in layer α, that is

$$b_{i\alpha} = 1 - \delta(0, k_i^{[\alpha]}). \tag{6.23}$$

The activity B_i of node i [21] is the number of layers in which node i is active, that is

$$B_i = \sum_{\alpha=1}^{M} b_{i\alpha}. \tag{6.24}$$

The layer activity $N^{[\alpha]}$ of layer α is the number of nodes connected in layer α. The correlation between the activities of two nodes can be measured using the *node pairwise multiplexity* Q_{ij} [23] given by

$$Q_{ij} = \frac{1}{M} \sum_{\alpha=1}^{M} b_{i\alpha} b_{j\alpha}. \tag{6.25}$$

The *layer pairwise multiplexity* $Q_{\alpha\alpha'}$ [21] measures instead the correlations among the layer in terms of the number of shared active nodes, that is

$$Q_{\alpha\alpha'} = \frac{1}{N} \sum_{i=1}^{N} b_{i\alpha} b_{i\alpha'}. \tag{6.26}$$

Using the layer pairwise multiplexity it is possible to construct a weighted network between the layers [21] where any two layers α and α' are connected by a link with weight $Q_{\alpha,\alpha'}$.

6.5   Null Models of Multiplex Networks

In the analysis of multiplex networks it is usually important to have null random-ized network models as a suitable benchmark for the multiplex network structure. Therefore the randomization procedures that can preserve or remove build-in corre-lations constitute fundamental tools for network scientists working on real biological networks.

Given the different types of correlations encoded in multiplex network struc-tures, different randomization procedures can be suggested. There are, neverthe-less, only three main procedures [3] (codes available at GitHub Repository https://github.com/ginestrab):

1. *Randomization of the replica nodes.* The same networks are kept intact in each layer but the label of the nodes in each layer is reshuffled, changing the identity of corresponding replica nodes [24] and removing inter-layer degree correlations.

2. *Independent randomization of each layer.* Each layer is randomized independently, keeping the same degree sequence. In this way the same inter-layer degree correlations are kept, but the intra-layer degree correlations and link overlap are removed. The algorithm applied to each single layer α proceeds as follows [3]:

    –   Consider two random links of layer α connected to four distinct nodes. We assume that the first link is connected to nodes i_1 and j_1 and that the second link is connected to nodes i_2, j_2.
    –   Swap the two links, substituting them with other two links connecting node i_1 to node j_2 and node i_2 to node j_1 only if the move is allowed. The move is allowed if none of the links (i_1, j_2) and (i_2, j_1) already exists in layer α.

    These swaps proceed until the network is completely randomized and basic structural properties of the network do not change by increasing the number of iterations.

3. *Randomization preserving multidegree sequence.* The multiplex network is randomized, keeping the same multidegree sequence [15]. In this way the link overlap and the inter-layer degree correlations are preserved.
    This generalized swap algorithm is defined as follows [3]:

    –   Consider two random multilinks of the same type $\vec{m} \neq \vec{0}$ connected to four distinct nodes of the network. Let us assume that the first multilink is connected to nodes i_1 and j_1 and the second multilink is connected to nodes i_2 and j_2.
    –   Swap the two multilinks substituting the original multilinks $\vec{m} \neq \vec{0}$ with two multilinks $\vec{m}$ connecting nodes i_1 and j_2 and nodes i_2 and j_1 if and only if the move is allowed. The move is allowed if the nodes i_1 and j_2 and the nodes i_2 and j_1 are not yet connected by any type of non-trivial multilink.

6.6   Multilink Community Detection

The literature on community detection of multiplex networks is very rich. One of the first papers on the subject [25] proposed a generalized modularity measure to characterize the persistence of community among different layers. However, this method extensively uses interlinks that we are not treating in this chapter. Alternative multi-layer network community detection algorithms include the consensus algorithms [5, 26], the PARAFAC tensor decomposition [27] and multilink community detection [12].

While most of the proposed multi-layer community detection methods associate a single community to a single node or to a single replica node, it is actually true that on the contrary in multiplex networks single nodes can belong to many communities. The *multilink community detection* algorithm overcomes this limitation by associating a single community assignment to each multilink. Therefore, this allows the assignment of several communities to each node of the multiplex network.

In particular, the multilink community detection algorithm [12] (code available at GitHub Repository https://github.com/ginestrab) starts by assigning a similarity to every pair of incident multilinks. This similarity measure gets larger as the Hamming distance between the two incident multilinks $\vec{m}$ and $\vec{m}'$ increases and the clustering is the local neighborhood of nodes of the two multilinks becomes greater. Starting from this similarity matrix, the algorithm performs a hierarchical clustering between multilinks. The resulting dendrogram is subsequently cut, corresponding to the optimal score function evaluating the best clustering among the multilinks. In this way the best multilink communities are found, while the study of the dendrogram provides also the hierarchical structure among their sub-communities.

Interestingly, this algorithm, when applied to biological as well as transportation and social networks, reveals that a node can have very different *layer activity* and *community activity*; that is, it is both possible that a node connected in several layers (high layer activity) belongs to many multilink communities (high community activity) as well as relatively few multilink communities (low community activity). Moreover, the multilink communities can have very different layer compositions. In particular, it is possible that a multilink community comprises multilinks including connections only in a few layers or alternatively multilinks including connections in a large variety of layers.

In Figure 6.5 we show the results related to the multilink communities of the *C. elegans* duplex networks comprising two layers: the layer of synaptic connections and the layer of gap junctions. These results show the significant heterogeneity in the multilink communities that can have very different sizes and can induce very heterogeneous community and layer activity. At the same time, the algorithm can reveal the similar role of neurons such as the AVAL and AVAR neurons.

6.7   Centrality Measures

Centrality measures have been extensively used in single-layer networks to characterize the relevance of the nodes. In multiplex networks, centrality measures can be

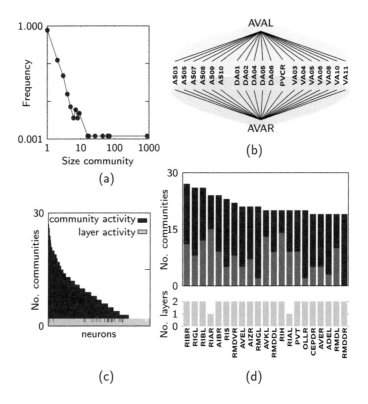

Figure 6.5 Characterization of the Multilink Community in a two-layer multiplex network of *C. elegans* formed by synaptic connections and gap junctions. (a) The distribution of the community sizes for the multiplex connectome of *C. elegans*. (b) The two most similar sub-communities contained in the largest multilink community. (c) Neurons ranked in decreasing order of their community activity. (d) Layer and community activity for the top-ranked neurons. (The contribution of communities with single multilinks is in pink, while the contributions of communities with more than one multilink is in blue.) Reprinted figure from [12]. © 2018 Mondragon et al.

defined in different ways (see, for instance, the versatility [28] defined in the next chapter). Here we focus on centrality measures that extend the PageRank centrality to multiplex networks without interlinks: the Multiplex PageRank [29], the Functional Multiplex PageRank [11] and the MultiRank [30]. The codes for these multi-layer centrality measures can be found in the GitHub Repository https://github.com/ginestrab.

6.7.1 Multiplex PageRank

In many situations we know that a node plays a central role on one network and therefore we are interested in exploring if its importance in the considered network can have a significant effect on another layer of the multiplex network. In this case, it is very useful to consider the Multiplex PageRank [29] that aims at capturing the influence of the centrality of a node in one layer (layer α) on the centrality of the same node on another layer (layer α'). In order to capture this effect, the PageRank centrality

$x_i^{[\alpha]}$ of each node i is measured in layer α. Subsequently, the Multiplex PageRank X_i of node i is measured in layer α'. The Multiplex PageRank is determined by the steady state of a random walker that starts from a given node and either hops to a neighbor node (by following a link of layer α') with probability μ or jumps to a random node of the network. However, both moves (hop and jump) can be biased by the centralities $\{x_i^{[\alpha]}\}_{i=1,2,...,N}$ of the nodes in layer α. In particular, when a node decides to hop it can choose the neighbor node i to hop to with a probability proportional to $x_i^{[\alpha]}$. Similarly, when jumping to a random node of the network it can choose the target node i proportionally to its centrality $x_i^{[\alpha]}$ in layer α. Therefore it is possible to distinguish between the following three non-trivial versions of the Multiplex PageRank, distinguished by the different values of the parameters q and n:

1. *Additive Multiplex PageRank.* In this case the hop is unbiased ($q = 0$) and the jump is biased ($n = 1$). In the Additive PageRank, a node derives an added benefit from being central in layer α.
2. *Multiplicative Multiplex PageRank.* In this case the hop is biased ($q = 1$) and the jump is unbiased ($n = 1$). In the Multiplicative Multiplex PageRank, every node derives an added benefit from being central in layer α, but this benefit is contingent on its connections in layer α'.
3. *Combined Multiplex PageRank.* In this case both the hop ($q = 1$) and the jump ($n = 1$) are biased. In the Combined Multiplex PageRank, the effect of layer α on the centrality of the nodes is a combined effect of the additive and multiplicative multiplex PageRank.

6.7.2 Functional Multiplex PageRank

The centrality of a node in a multiplex network can be strongly affected by the relevance attributed to different types of links. However, in many cases we do not have access to information sufficient to evaluate which pattern of connection is more important in a multiplex network. In this case, it is useful to consider the Functional Multiplex PageRank algorithm, which associates to each different type of non-trivial multilink $\vec{m}$ an *influence* $z^{\vec{m}}$ that is a free parameter that takes any possible value between 0 and 1.

In the traditional single-network PageRank algorithm, a node increases its centrality if very central nodes point already to it. In the multiplex network version of the PageRank, called Functional Multiplex PageRank [11], it is assumed that the centrality of a node increases if already very central nodes point to it through very influential multilinks $\vec{m}$. As for the original PageRank algorithm so also for the Functional Multiplex PageRank, the centrality of a node is determined in terms of a random walker that starts from a node and either hops to a neighbor node or jumps to a random node of the network. In particular, if the random walker that determines the Functional Multiplex PageRank is on a given node, it can hop to a node connected to the initial one in every layer; however, the probability of hopping to a neighbor connected via a multilink $\vec{m}$ is proportional to the influence $z^{\vec{m}}$ of the multilink. Finally, the Functional Multiplex PageRank $X_i(\mathbf{z})$ of a node i is the steady-state probability that the random walker is at node i, given the set of influences $\mathbf{z} = \{z^{\vec{m}}|\vec{m} \neq \vec{0}\}$ attributed to each non-trivial multilink.

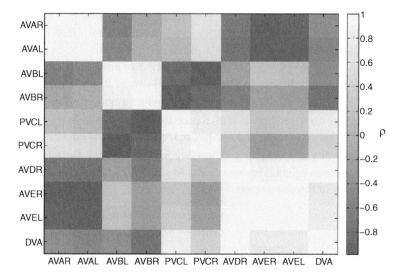

Figure 6.6 Correlations among the Functional Multiplex PageRank of top-ranked neurons in the two-layer connectome of C. *elegans* formed by synaptic connections and gap junction. These correlations are higher among neurons of the same cell type, the showing that the Functional Multiplex PageRank can reveal important information about the centrality and multi-layer role of the nodes in the multiplex network. Reprinted from [11] under the Creative Commons Attribution 3.0 License; Copyright © EPLA, 2016.

The peculiar property of the Functional Multiplex PageRank is that the centrality of a node in the multiplex network is described by an entire function $X_i(\mathbf{z})$ of the influences of the multilinks $\mathbf{z}$. This property allows us to characterize the centrality of a node with respect to the relevance attributed to different patterns of connection. Interestingly, this can be used to measure the correlations existing between the Functional Multiplex PageRank of different nodes, characterizing in this way nodes that benefit from the same type of connections.

Finally, from the Functional Multiplex PageRank of the nodes $\{X_i(\mathbf{z})\}_{i=1,2,...,N}$ it is also possible to extract a global ranking $\{\hat{X}_i\}_{i=1,2,...,N}$ of the nodes by associating to each node i of the multiplex network the Absolute Functional Multiplex PageRank given either by its average or its maximum Functional Multiplex PageRank.

This algorithm has been shown to reveal important properties of biological networks and infrastructure networks. In particular, the application of this algorithm to the connectome of C. *elegans* has revealed that the correlations of the Functional Multiplex PageRank of top-ranked neurons can clearly detect neurons with the same cell type (Figure 6.6).

The Functional Multiplex PageRank has great flexibility since it does not assign any predefined influences to multilinks and is able to capture important properties of multiplex networks with fine detail. However, its use is limited to multiplex networks with few layers as the number of multilinks increases rapidly with the number of layers and the computational cost becomes significant. In order to address this limitation, the MultiRank algorithm has been recently proposed. The MultiRank is

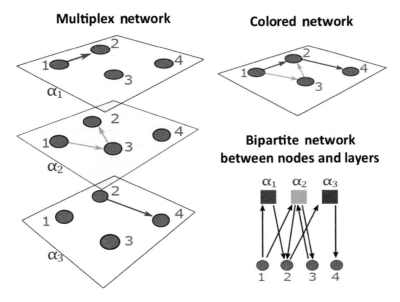

Figure 6.7 Schematic representation of a multiplex network and its description in terms of a colored network and a bipartite network between nodes and layers. This representation of a multiplex network is used by the MultiRank algorithm to provide a simultaneous centrality measure of nodes and layers of a multiplex network. Reprinted from [30]. Copyright © 2017, Oxford University Press.

able to efficiently calculate the centrality of nodes in multiplex networks with many (hundreds) of layers.

6.7.3   MultiRank

The MultiRank algorithm exploits a description of the multiplex network in terms of a single-colored network describing the interactions of different natures existing within each layer and a bipartite network formed by nodes and layers, and indicates which node is active in which layer (Figure 6.7).

Using this description, the MultiRank algorithm ranks simultaneously the nodes and the layers of a multiplex network. The centrality $z^{[\alpha]}$ of a layer is called the *layer influence*. The centrality X_i of each node of the multiplex network can be evaluated using a biased random walker.

In particular, the centrality of a node X_i is the steady-state probability of a random walker that starts from a node and can either hop on a neighbor in layer α with probability proportional to $z^{[\alpha]}$ or can jump to a random node of the multiplex network. However, the influences $z^{[\alpha]}$ are not free to be assigned a priori, or to take any possible allowed value as in the Functional Multiplex PageRank; rather, they are determined by imposing that layers have larger influence if very central nodes are already active on it (connected). Therefore, the MultiRank is obtained by coupling the equation for the centrality of the nodes $\{X_i\}_{i=1,2,...,N}$ to the equation for the influences of the layer $\{z^{[\alpha]}\}_{\alpha=1,2,...,M}$. In this way, the MultiRank can very efficiently provide the centrality of nodes and layers in multiplex networks with up to hundreds of layers.

6.8   Conclusions

Multiplex networks are the most simple and fundamental multi-layer networks. They are formed by N nodes connected in M layers, with each layer characterizing the set of interactions with the same connotation. As such, multiplex networks are very suitable structures to encode data on molecular networks. In fact, multiplex networks can be used to combine information coming from experiments conducted under different conditions or on different tissues, like different gene-expression experiments or to combine information coming from different types of experiments, like gene co-expression and co-regulation of genes through the same transcription factor.

The representation of biological datasets in terms of a multiplex network opens new perspectives for quantify the information encoded in them. In fact, multiplex networks encode information in their highly correlated structure, including link overlap, inter-layer degree correlations and heterogeneous activity of the nodes. Interestingly, these correlations can be measured in real networks and the obtained result can be compared to suitable null models designed to remove or preserve one or more of these correlations.

The very complex pattern of connections that can exist among any two nodes in a multiplex network is fully captured by the multilinks that can be used as motifs to extract relevant information from a dataset. Moreover, the multidegree evaluating the number of multilinks incident to each node is a fundamental measure for capturing the role of the nodes in a multiplex network. Multilinks can also be used to capture the community detection of multiplex networks using the Multilink Community Detection associating to each multilink a given community assignment. Interestingly, in this perspective it can be observed that the communities of a multiplex network can either belong to just a few layers or to many layers. Additionally, there is also very strong heterogeneity in the role of each node in the mesoscale structure described by the multilink communities. In fact, a node active in many layers can belong to many or just a few multilinks communities.

Finally in a number of relevant scenarios it is important to characterize the centrality of nodes in multi-layer networks. In this chapter we have presented a series of centrality algorithms that exploit the built-in correlations existing in the multiplex network. The Multiplex PageRank measures the centrality of a node in one layer by using a PageRank algorithm biased by the centrality of the nodes according to a PageRank algorithm performed on another layer of the multiplex network. Therefore the multiplex PageRank is sensible to the inter-layer degree correlations among the nodes. The Functional Multiplex PageRank instead depends a set of *influences* associated to the multilinks of the multiplex network. In this framework the centrality of a node is no longer a single number but is a function of the influences attributed to the multilinks. Therefore it is possible to measure correlations among the Functional Multiplex PageRank and explore for each nodes which are the type of multilinks from which it benefits the most. The Functional Multiplex PageRank is very informative. However, its use is limited to multiplex networks with few layers. In order to capture the centrality of the nodes in large multiplex networks comprising many layers, the MultiRank proposes to simultaneously rank both nodes and layers of a given multiplex network and can be used efficiently on multiplex networks with hundreds of layers.

In conclusion, in this chapter we have defined multiplex networks and we have presented how to perform a multiplex network analysis. We have discussed several algorithms to extract relevant information from them, including the use of multilinks as motifs, the multilink community detection algorithm and a number of centrality measures, and we have illustrated the relevance of these algorithms for characterizing the network of networks in the cell.

References

[1] Barabási A-L, Gulbahce N, Loscalzo J. Network medicine: a network-based approach to human disease. *Nature Rev Gen.* 2011;12:56.

[2] Menche J, Sharma A, Kitsak M, et al. Uncovering disease–disease relationships through the incomplete interactome. *Science.* 2015;347:1257601.

[3] Bianconi G. *Multi-layer Networks: Structure and Function.* Oxford: Oxford University Press, 2018.

[4] Boccaletti S, Bianconi G, Criado R, et al. The structure and dynamics of multi-layer networks. *Phys Rep.* 2014;544:1.

[5] Cantini L, Medico E, Fortunato S, Caselle M. Detection of gene communities in multi-networks reveals cancer drivers. *Sci Rep.* 2015;5:17386.

[6] Li W, Liu C-C, Zhang T, et al. Integrative analysis of many weighted co-expression networks using tensor computation. *PLoS Comp Bio.* 2011;7:e1001106.

[7] Menichetti G, Remondini D, Bianconi G. Correlations between weights and overlap in ensembles of weighted multiplex networks. *Phys Rev E.* 2014;90(6):062817.

[8] Bullmore E, Sporns O. Complex brain networks: graph theoretical analysis of structural and functional systems. *Nat Rev Neurosci.* 2009;10(3):186–198.

[9] Sporns O. *Networks of the Brain.* Cambridge, MA: MIT Press, 2010.

[10] Bentley B, Branicky R, Barnes CL, et al. The multi-layer connectome of *Caenorhabditis elegans.* *PLoS Comp Bio.* 2016;12:e1005283.

[11] Iacovacci J, Rahmede C, Arenas A, Bianconi G. Functional multiplex PageRank. *EPL.* 2016;116:28004.

[12] Mondragon RJ, Iacovacci J, Bianconi G. Multilink communities of multiplex networks. *PLoS One.* 2018;13(3):e0193821.

[13] Bassett DS, Wymbs NF, Porter MA, et al. Dynamic reconfiguration of human brain networks during learning. *Proc Nat Acad Sci USA.* 2011;108:7641.

[14] Srivastava A, Kumar S, Ramaswamy R. Two-layer modular analysis of gene and protein networks in breast cancer. *BMC Syst Biol.* 2014;8(1):81.

[15] Bianconi G. Statistical mechanics of multiplex networks: entropy and overlap. *Phys Rev E.* 2013;87:062806.

[16] Radicchi F, Bianconi G. Redundant interdependencies boost the robustness of multiplex networks. *Phys Rev X.* 2017;7:011013.

[17] Cellai D, Dorogovtsev SN, Bianconi G. Message passing theory for percolation models on multiplex networks with link overlap. *Phys Rev E*. 2016;94:032301.

[18] Musmeci N, Nicosia V, Aste T, Di Matteo T, Latora V. The multiplex dependency structure of financial markets. *Complexity*, 2017: 9586064.

[19] Menichetti G, Remondini D, Panzarasa P, Mondragón RJ, Bianconi G. Weighted multiplex networks. *PLoS One*. 2014;9:e97857.

[20] Nicosia V, Bianconi G, Latora V, Barthélemy M. Growing multiplex networks. *Phys Rev Lett*. 2013;111:058701.

[21] Nicosia V, Latora V. Measuring and modelling correlations in multiplex networks. *Phys Rev E*. 2015;2:032805.

[22] Cellai D, Bianconi G. Multiplex networks with heterogeneous activities of the nodes. *Phys Rev E*. 2016;93:032302.

[23] Criado R, Flores J, García del Amo A, Gómez-Gardeñes J, Romance M. A mathematical model for networks with structures in the mesoscale. *Int J Comput Math*. 2012;89:291.

[24] Min B, Yi SD, Lee K-M, Goh K-I. Network robustness of multiplex networks with interlayer degree correlations. *Phys Rev E*. 2014;89:042811.

[25] Mucha PJ, Richardson T, Macon K, Porter MA, Onnela JP. Community structure in time-dependent, multiscale, and multiplex networks. *Science*. 2010;328:876.

[26] Lancichinetti A, Fortunato S. Consensus clustering in complex networks. *Sci Rep*. 2012;2.

[27] Gauvin L, Panisson A, Cattuto C. Detecting the community structure and activity patterns of temporal networks: a non-negative tensor factorization approach. *PLoS One*. 2014;9:e86028.

[28] De Domenico M, Solé-Ribalta A, Omodei E, et al. Ranking in interconnected multi-layer networks reveals versatile nodes. *Nature Comm*. 2015;6:6868.

[29] Halu A, Mondragón RJ, Panzarasa P, Bianconi G. Multiplex PageRank. *PLoS One*. 2013;8:e78293.

[30] Rahmede C, Iacovacci J, Arenas A, Bianconi G. Centralities of nodes and influences of layers in large multiplex networks. *arXiv preprint. arXiv:1703.05833*, 2017.

7 Existing Tools for Analysis of Multi-layer Networks

Massimo Stella and Manlio De Domenico

7.1 Introduction

In this chapter, we review several structural metrics for multi-layer networks, highlighting differences against definitions for single-layer networks. We start by extending the concepts of network neighbours and network paths to the multi-layer structure. We then proceed by considering how the presence of multiple layers requires more general definitions of local and global connectivity, centrality and community structure. We also review structural features arising from multi-layer structures that have no single-layer counterpart, such as interlayer correlations [1–5], structural reducibility [6] and layer clustering [6, 7]. Notice that the algorithms for multi-layer community discovery [8, 9] briefly introduced here link together structure and dynamics of multi-layer networks and are reviewed in more depth in Chapter 8.

7.2 Multi-layer Neighbours, Links and Paths

Let us consider a multi-layer network [1, 10, 11] with L layers and N nodes represented by the multi-layer adjacency tensor $M_{j,\beta}^{i,\alpha}$, with $i, j \in 1, 2, \ldots, N$ and $\alpha, \beta \in 1, 2, \ldots, L$. Then the neighbours of the generic node i are all the nodes connected to i across all layers. Notice that even if two nodes i and j might be disconnected in one layer α, they might be adjacent on another layer β or be interconnected across layers. As in single-layer networks [12], neighbours of node i are classified in terms of the length of the paths connecting them to i. In the rest of this chapter, using 'neighbours' with no further attribute indicates adjacent nodes. The presence of several network layers can drastically alter the neighbourhood structure of a given node. An example is reported in Figure 7.1: even though nodes 1 and 5 are disconnected in the left-most layer, there is a path connecting 1 and 5 in the central layer.

Let us briefly discuss the concept of paths within the framework of multi-layer networks. As in single-layer networks [12], we define an undirected path between nodes i and j as a set of network links $p = \{(i, \alpha, k, \alpha), (k, \alpha, k, \beta), \ldots, (h, \gamma, j, \gamma)\}$ where

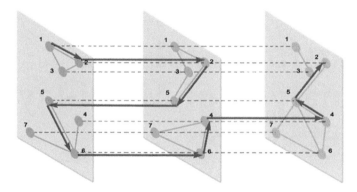

Figure 7.1 A node-aligned multi-layer network with $N = 7$ nodes and $L = 3$ layers. The red arrows represent a path connecting node 1 on the left-most layer to node 2 on the right-most layer. Notice that the path considers both links between nodes in a given layer (intra-layer links) and connections between replicas of nodes in different layers (inter-layer links). Figure from De Domenico et al. 2014 [13].

directionality is not important and the notation is such that (i, α, k, α) indicates an *intra-layer* link between nodes i and k in layer α, while (k, α, k, β) indicates an *inter-layer* link between replicas of node k in layers α and β (see Figure 7.1 for an example). The path length is then defined as the cardinality $|p|$ of p – that is, the number of links in a given path. The shortest path length is the cardinality of the path with the fewest inter- and intra-links connecting nodes i and j. Notice that the possibility of transitioning between layers can dramatically change the shortest path length between nodes (also called geodesic distance [12]): two nodes could be disconnected on layer α, so that their distance would be $d_\alpha(i, j) = \infty$, but adjacent in another layer; consequently their distance on the multi-layer structure would be $d_M(i, j) = 1$. A multi-layer network and its aggregated single-layer counterpart display the same distances between nodes only when inter-layer links are not explicitly considered, as in edge-coloured graphs [1, 13].

7.2.1 Viable Clusters and Mutually Largest Connected Components

In single-layer networks [12], a *connected component* is a subgraph in which there is at least one path connecting any two nodes. For networks of finite size, the connected component with the largest number of nodes is called the *largest connected component*. For networks of infinite size, such as theoretical models, the largest connected component is called a *giant component*. A network is connected if it coincides with its largest connected component.

In multi-layer networks, the presence of different layers of interactions allows us to consider several definitions of connectedness [1, 10, 11, 13–16]. The simplest approach for node-aligned multi-layer networks is to apply the single-layer notion of connectedness to the aggregate network [1], so that a multi-layer network is fully connected if its aggregate counterpart is fully connected. According to this definition, in a connected multi-layer network there exists at least one multi-layer path connecting any two nodes.

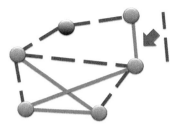

Figure 7.2   Example of viability in an edge-coloured network with $N = 6$ nodes and $L = 2$ layers. Red nodes are the intersection of the largest connected components across layers. Viability requires all nodes to be connected among each other in every individual layer and paths combining different colours do not count. The intersection of connected components is not a viable cluster: There are no paths of dashed links only connecting the node in the upper-right corner. In order to make the nodes a viable cluster, one dashed link must be added. Figure readapted from Stella et al [17]. Copyright © 2018, Springer Nature

Connectedness can be imposed on individual layers rather than on the aggregate network. For node-aligned multiplex networks without explicit interlinks, this requirement leads to the definition of a viable cluster [16]: a set of nodes that are mutually connected on each individual layer. The notion of viable cluster reduces to the one of largest connected component in single-layer networks. In edge-coloured graphs the viable cluster is always smaller than the intersection of the largest connected components of individual layers [17] – see Figure 7.2 for an example.

Viable clusters are related to mutually connected components in interdependent networks [14, 15]. Each node i on layer α is in the mutually connected component if: (1) it has at least one neighbour j on α that is connected to the mutually connected component and (2) if all the replicated nodes i on other layers are also in the mutually connected component. The second requirement about interconnectedness makes the mutually connected component a more restrictive definition compared to the one of connectedness provided through the aggregate network. In the case of full interdependency, the definition of the largest mutually connected component corresponds to one of the largest viable cluster [16].

Notice that the way nodes are connected between layers and the presence of viable clusters or mutually connected components greatly affects the robustness properties of multi-layer networks to progressive disruption [13, 17, 18] and resilience to cascading failure [14, 19].

7.3   Centrality Measures in Multi-layer Networks

Finding the most central nodes in complex networks is fundamental in a variety of real-world scenarios [12], such as finding the most fragile agents in cascades of failure [14] or pivotal disease spreaders [20, 21] in epidemics. Centrality strongly depends on the considered network process so that a wide variety of network centrality measures has been suggested in single-layer networks over the years (for a review from the computational social sciences, see [22]). Generalizations to multi-layer networks are not always straightforward, since nodes peripheral in one layer might be extremely central in another one [1, 10, 11]. Although attempts at producing weighted averages of centralities [23] across single layers have been successfully used in multi-layer network analysis [24], many novel measures harnessing the multi-layer structure have been suggested over the last few years. Because of space constraints, our review cannot be exhaustive, so we refer the interested reader to consult additional references for multiplex [25] and multi-layer [1, 10, 11] networks.

7.3.1   Node Activity

In multi-layer networks, one node might be present on one layer but absent (or disconnected) on another one. In node-aligned multiplex networks this information can be encoded within a node–activity vector [4]:

$$\mathbf{b}_i = (b_i(1), b_i(2), \ldots, b_i(L)) \tag{7.1}$$

where $b_i(\alpha) = 0$ if $k_i^{(\alpha)} = 0$ and $b_i(\alpha) = 1$ otherwise, for $\alpha \in 1, \ldots, L$. The node activity B_i is then defined as the sum of the $b_i(\alpha)$ across layers. The distribution of node activities provides a compact representation of the involvement of nodes across different layers [4, 25].

7.3.2   Multidegree

Counting the links of a given node represents the simplest indicator of node importance at a local level, a measure called degree centrality in single-layer networks [12]. In multi-layer networks, *multidegree centrality* [1] is the total number of links k_i in which a node i is involved across all layers, and it can be computed through the multi-layer adjacency tensor:

$$k_i = \sum_{\alpha,\beta=1}^{L} \sum_{j=1}^{N} M_{j\beta}^{i\alpha} \tag{7.2}$$

Notice that the multidegree k_i counts both inter- and intra-layer links. In case inter-layer links are not explicitly considered, as in edge-coloured graphs, it can be more convenient to consider the degrees coming from individual layers/colours:

$$k_i^{(\alpha)} = \sum_{j=1}^{N} A_j^i(\alpha) \tag{7.3}$$

where k_i^α is the degree of node i coming from intra-layer links on layer α and $A_j^i(\alpha)$ represents the adjacency tensor of layer α. When inter-layer links are not present, the multidegree coincides with the sum of intra-layer degrees $K_i = \sum_\alpha k_i^{(\alpha)}$, which is also called the *overlapping degree* in the multiplex networks literature [3, 4, 25, 26]. Notice that even in multiplex networks the multidegree of i across layers does not always coincide with the degree of i in the aggregated network as links might overlap across layers [26].

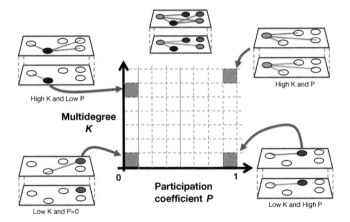

Figure 7.3   Visual explanation of a multiplex cartography. Nodes are mapped into points of a 2D map displaying the participation coefficient on the x axis and multidegree on the y axis. Points can be clustered in a 2D grid for easier visualization [4]. Figure from Stella et al. [28].

Network Cartography

Multidegree centrality k_i provides global information on the amount of interactions node i engages with. However, multidegree is not informative on the way i's links are distributed across layers. In multiplex networks, the participation coefficient [3] P_i quantifies how uniformly node i distributes their own intralayer links across layers:

$$P_i = \frac{L}{L-1} \left[1 - \sum_{\alpha=1}^{L} \left(\frac{k_i^{(\alpha)}}{K_i} \right)^2 \right] \tag{7.4}$$

P_i ranges between 0 (for nodes that concentrate all their connections in one layer only) and 1 (for nodes that distribute connections over all L layers uniformly).

The degree centrality and the participation coefficient of nodes in a multiplex network represent the coordinates of the so-called *multiplex cartography* [3], that is a 2D map useful for detecting central nodes (Figure 7.3). The concept of network cartography was originally introduced for community analysis in single-layer networks [27], and it was later generalized to node centrality in multiplex networks [3]. Nodes with higher multidegree centrality and higher participation coefficient tend to have more interactions present across all layers [3, 4], thus playing a more central role in processes such as epidemic spreading on multiple layers [20, 21].

7.3.3   Multi-layer Versatility

In multi-layer networks, nodes crucial for a given dynamics might not be central across all layers. Consider the case in which two distinct layers have only one node in common: any information flow between the layers will have to pass through the common node, independently of its centrality in the layers. Hence that node will be highly central for the considered process. Multi-layer versatility [1, 29] quantifies how important nodes are for diffusive processes such as information flow or spreading. Since versatility is strongly connected to diffusive processes, we hereby briefly review different types of versatility measures but refer the reader to Chapter 8.

Eigenvector Versatility

In single-layer networks, eigenvector centrality is based on an iterative procedure giving to each node a centrality score that is the weighted sum of the scores of its neighbours [12]. Therefore, the eigenvector centrality gives high centrality scores to vertices that are connected to many other well-connected vertices.

As for single-layer networks, eigenvector versatility can be defined as the solution of an eigenvalue problem given by the multi-layer adjacency tensor:

$$M_{j\beta}^{i\alpha}\Theta_{i\alpha} = \lambda_1\Theta_{j\beta} \tag{7.5}$$

where Einstein summation is considered for the sake of simplicity; that is, there is summation over repeated indices. The problem of finding the eigenvector centrality consists in computing $\Theta_{j\beta} = \lambda_1^{-1}M_{j\beta}^{i\alpha}\Theta_{i\alpha}$, which represents the multi-layer generalization of Bonacich's eigenvector centrality per node per layer [1, 29]. Eigenvector versatility can then be condensed across layers by summing up the scores of a node across all layers $\theta_i = \sum_\alpha \Theta_{i\alpha}$. Notice that the summation across layers appears naturally through the tensorial formalism [1, 29], at variance with other measures [23, 30] based on specific prescriptions on how to combine the centrality in different layers.

Katz Versatility

Katz centrality was introduced in the field of social network analysis for 'evaluating status in a manner free from the deficiency of popularity contest procedures' [31]. This metric attributes centrality to a given node by considering contributions coming from its adjacent and more distant neighbours. Contributions coming from more distant neighbours are dampened by a factor a. Analogously to the eigenvector versatility, also Katz versatility $\Phi_{j\beta}$ can be defined as the solution of a tensorial equation $\Phi_{j\beta} = aM_{j\beta}^{i\alpha}\Phi_{i\alpha} + bu_{j\beta}$ [1, 29]. In formulas, the Katz centrality for a multi-layer network is:

$$\Phi_{j\beta} = [(\delta - aM)^{-1}]_{j\beta}^{i\alpha}U_{i\alpha} \tag{7.6}$$

where $\delta_{j\beta}^{i\alpha} = \delta_j^i\delta_\beta^\alpha$, the dampening factor a has to be smaller than the reciprocal of the absolute value of the largest eigenvalue of the multi-layer tensor $M_{j\beta}^{i\alpha}$ and b is a constant usually equal to 1. As for the eigenvector versatility, the overall Katz centrality of a node is the sum of centrality scores $\Phi_{j\beta}$ across layers [1, 29].

HITS Versatility

In single-layer networks, the hyperlink-induced topic search, or HITS, centrality was originally introduced for web page rating according to their authority (i.e. their content) and their hub value (i.e. the value of their links to other web pages) [32]. Nodes pointing to many other nodes are considered hubs, while nodes receiving links by many hubs are considered authorities. The problem of computing hub and authority values for nodes translates into two coupled eigenvalue problems whose multi-layer counterpart is [1, 29]:

$$\begin{cases} (MM^T)^{i\alpha}_{j\beta}\Gamma_{i\alpha} = \lambda_1\Gamma_{j\beta} \\ (M^TM)^{i\alpha}_{j\beta}\Upsilon_{i\alpha} = \lambda_1\Upsilon_{j\beta} \end{cases} \tag{7.7}$$

where ^{T} indicates the transpose operation, $\Gamma_{i\alpha}$ ($\Upsilon_{i\alpha}$) is the hub (authority) centrality score of node i on layer α. The overall HITS versatilities are then obtained by summing across layers. Notice that for undirected networks with no interlayer links, the HITS versatilities coincide with the eigenvector versatility.

PageRank Versatility

In single-layer networks, PageRank centrality [33] indicates the probability that a random walker visits a given node by: (1) traversing node links and (2) teleporting through nodes at random. For multi-layer networks [1, 29], the dynamics of the random walker is regulated by a transition tensor, expressing the probability of traversing link (i, α, j, β):

$$R^{i\alpha}_{j\beta} = rT^{i\alpha}_{j\beta} + \frac{(1-r)}{NL}u^{i\alpha}_{j\beta} \tag{7.8}$$

where r is a constant expressing the probability for the random walker to teleport. In the Google Search algorithm, r is equal to 0.85 [33]. If we denote by $\Omega_{i\alpha}$ the eigentensor of $R^{i\alpha}_{j\beta}$, then the PageRank versatility ω_i of a node is obtained by summing across layers $\sum_\alpha \Omega_{i\alpha}$. Other definitions of multi-layer PageRank have been proposed in the context of multiplex networks [30, 34, 35]. Multi-layer PageRank is more powerful than its single-layer counterpart in a wide variety of applications, including importance [34], tendency for there to be passenger traffic congestion at airports [29] and identifying interdisciplinary researchers in co-citation networks [36].

7.4   Correlations in Multi-layer Networks

Intra-layer connections can be organized in correlated ways among layers. Several attempts have been made to quantify these inter-layer dependencies [1, 3, 4, 37]. Let us briefly review quantitative measures of correlation relying on link assortment across layers.

7.4.1   Link Overlap and Multiplexity

A simple measure of correlation among links in node-aligned multi-layer networks is *link overlap* [1, 26], the number of links being simultaneously present on several layers at once between the same couple of nodes. Correlations influencing link overlap can be measured via correlation metrics, like the conditional probability $P(A^i_j(\alpha)|A^i_j(\beta))$

of finding a link at layer α given the presence of a link between the same nodes in layer β:

$$P(A^i_j(\alpha)|A^i_j(\beta)) = \frac{\sum_{ij} A^i_j(\alpha) A^i_j(\beta)}{\sum_{ij} A^i_j(\beta)} \tag{7.9}$$

However, such an approach does not quantify the contribution that heavy-tailed degree distributions might have in boosting or altering correlations on link overlap. The metric called *multiplexity* $M_{\alpha\beta}$ allows us to quantify pairwise link overlap correlations between layers α and β by considering a specific null model [5]:

$$M_{\alpha\beta} = \frac{2 \sum_{i \neq j} \min\{A^i_j(\alpha), A^i_j(\alpha)\}}{L^{[\alpha]}_{TOT} + L^{[\beta]}_{TOT}}, \quad \mu_{\alpha\beta} = \frac{M_{\alpha\beta} - \langle M_{\alpha\beta} \rangle}{1 - \langle M_{\alpha\beta} \rangle} \tag{7.10}$$

where $L^{[\cdot]}_{TOT}$ is the number of edges in a given layer and $\mu_{\alpha\beta}$ is the normalized multiplexity, which is rescaled compared to a reference null model value $\langle M_{\alpha\beta} \rangle$. When applied to the International Trade Network [5], where nodes are countries and layers indicate traded commodities, strong inter-layer link correlations were hugely reduced once the heavy-tailed empirical degree distribution was considered. Hence, with the proper choice of reference null model (e.g. a configuration model fixing degree but randomizing links), multiplexity can discriminate between genuine link correlations and spurious effects due to inter-layer degree correlations.

7.4.2   Degree Correlations

Degree correlations of the same node across different layers can be identified by computing the analogue of the nearest neighbour average degree $k_{nn}(k)$ of nodes with degree k in single-layer networks. For node-aligned multi-layer networks, one can define [1, 25]:

$$\overline{k^{(\beta)}}(k^{(\alpha)}) = \sum_{k^{(\beta)}} k^{(\beta)} P(k^{(\beta)}|k^{(\alpha)}) \tag{7.11}$$

which is the average degree at layer β of a node with degree $k^{(\alpha)}$ on layer α. Increasing (decreasing) trends of $\overline{k^{(\beta)}}(k^{(\alpha)})$ indicate positive (negative) degree correlations among layers α and β.

7.4.3   Triadic Closure and Clustering Coefficient

In single-layer networks, the clustering coefficient measures the presence and strength of triadic closure, that is the tendency for three connected nodes (a triplet) to form a triangle (a closed triplet) [12]. The local clustering coefficient of node i measures how likely it is for two neighbours of i to be connected to each other:

$$C_i = \frac{\sum_{j,m \neq i} A^i_j(\alpha) A^j_m(\alpha) A^m_i(\alpha)}{\sum_{j,m \neq i} A^i_j(\alpha) A^m_i(\alpha)} \tag{7.12}$$

AAA   AACAC   ACAAC   ACACA   ACACAC

Figure 7.4 Some examples of possible 3-cycles on a multi-layer structure. The orange node is the starting point of the cycle; intra-layer links are represented as solid lines and inter-layer links are dotted lines. The green line represents the second intra-layer link in the cycle. Figure from Cozzo et al. [38]. © 2015 IOP Publishing Ltd and Deutsche Physikalische Gesellschaft.

In multi-layer networks there are multiple ways of considering closed triangles across layers – see Figure 7.4 for some possible examples. Consequently, there are several different ways to generalize the notion of clustering to multi-layer network structures. For instance, in node-aligned multiplex networks, a triangle can be formed by considering two intra-layer links on layer α and one intra-layer link on β among three nodes, so that a possible generalization of local clustering becomes [3]:

$$C_{i,1} = \frac{\sum_{\alpha} \sum_{\beta \neq \alpha} \sum_{j,m \neq i} A_j^i(\alpha) A_m^j(\beta) A_i^m(\alpha)}{(L-1) \sum_{\alpha} \sum_{j,m \neq i} A_j^i(\alpha) A_i^m(\alpha)} \tag{7.13}$$

Notice that this definition averages above all possible couples of layers and the denominator considers also terms $j = m$, so that the maximum value of local clustering coefficient is $(N-1)/(N-2)$. Other attempts have been made to have measures of clustering with additional features such as: (1) naturally reducing to the single-layer definition; (2) being bounded between 0 and 1; (3) defining for node–layer pairs; and (4) defining also for non-node-aligned networks. All these features were encompassed in a recently suggested definition of multi-layer clustering based on cycles [38] and random walks:

$$C^* = \frac{\sum_i t_{*,i}}{\sum_i d_{*,i}} \tag{7.14}$$

counting the number of 3-cycles of a given type $*$ (e.g. cycles being in one layer only, cycles jumping across two layers and so on) going through node i normalized by a factor $d_{*,i}$.

The analysis of multi-layer clustering found a strong tendency for social networks to promote link redundancy by displaying triadic closure on every layer, an effect that would be otherwise unnoticeable by considering aggregate networks only [38]. Further investigation of multiplex configuration models [39] highlighted that degree correlations have a strong influence on the number of cycles a node participates in. This underlines the need for a comparison of clustering coefficient against suitable null models in real-world networks.

7.5 Community Discovery

A prominent problem in network science is the detection of densely connected groups of nodes known as communities [40]. For single-layer networks, a variety of methods has been developed in the last 20 years, either directly maximizing a given quality function of the detected community structure [41] or rather adopting information-theoretic tools [42] or spectral features [43] of complex networks.

The multi-layer structure challenges the concept of community in terms of tightly connected nodes, since nodes adjacent in one layer might be disconnected in another one and, in general, the mesoscale community organization of nodes in one layer might dramatically differ across other layers [8, 37]. This difference indicates the importance of taking into account interlayer dependencies and correlations within the definition of multi-layer communities.

7.5.1   Multi-slice Modularity Maximization

One of the most popular heuristics for community detection in single-layer networks is modularity maximization [40, 41]. Modularity is a metric indicating the extent to which the distribution of links across and within a given set of communities differs compared to what one would expect in a suitable network null model. Hence, finding a network community becomes an optimization problem of the modularity Q. A generalization of modularity maximization for multi-layer networks is the multi-resolution method [8], which employs a specific analytical generalization of modularity to multi-layer networks. Making use of the tensorial notation [1], the multi-layer modularity is written as:

$$Q_M \propto S_{i\alpha}^a (M_{j\beta}^{i\alpha} - P_{j\beta}^{i\alpha}) S_a^{j\beta} \tag{7.15}$$

where $M_{j\beta}^{i\alpha}$ is the multi-layer adjacency tensor, $P_{j\beta}^{i\alpha}$ is the tensor encoding an appropriate null model for the observed network structure and $S_a^{i\alpha}$ is defined to be 1 when node i in layer α belongs to a given community labelled by a and 0 otherwise. The algorithm then proceeds with the same heuristics of single-layer networks for detecting optima in the landscape of modularity. For a more thorough discussion of the dynamical interpretation of Q_m, see Chapter 8.

7.5.2   InfoMAP

Spectral clustering in single-layer networks relies on the intuitive idea that a random walker tends to remain confined preferentially within a given network community rather than jumping across communities [13, 43, 44]. By defining a map equation for both single- [42] and multi-layer [9] network structures, encapsulating specific types of random walkers on networks, the *InfoMAP* algorithm allows us to detect multi-layer communities. Since the map equation is a flow-based method operating on the network dynamics, we refer the interested reader to Chapter 8 for further details.

7.5.3   Non-negative Matrix Factorization for Temporal Networks

Temporal networks can be considered as multiplex networks with layers having a specific ordering and representing time [45]. Non-negative matrix factorization has been suggested for detecting community structure in temporal networks [45], represented as a three-dimensional tensor T_{ij}^τ. Kruskal decomposition is applied to T_{ij}^τ in order to assign each node to a community across nodes. Notice that this methodology assumes as stationary the interdependences across replica nodes over time.

7.5.4   Structural Reducibility

The effort of considering more network layers does not always provide additional information about interactions among agents: consider a node-aligned multi-layer or multiplex network in which all layers are copies of a single one. More in general, while individual layers can differ from each other in terms of their topology, they can also display rather high link overlap and thus contain redundant topological patterns [26, 46]. This naturally leads to the following question: what would be the optimal number of layers to consider in a trade-off between preserving the most topological information available and the least effort in terms of different layers to be considered?

To this end structural reducibility analysis was recently proposed [6] as a technique for quantifying how different a multiplex network is from its aggregate when its layers are kept either as distinct or rather gradually aggregated. Structural reducibility encodes the information available in the network structure through the Von Neumann entropy h [47], which is based on the spectrum $\{\lambda_i\}_i$ of the degree-normalized[1] Laplacian $L = (D - A)/2K$:

$$h = -\mathrm{Tr}(L\log_2 L) = -\sum_i \lambda_i \log_2 \lambda_i \qquad (7.16)$$

where Tr indicates the trace operator on matrices. The Von Neumann entropy h_A of the aggregated network is used as a reference value. Differences between h_A and the Von Neumann entropy of multiplex structure are then used for estimating the loss of information relative to aggregating the multiplex structure [6]. The Von Neumann entropy h_M of the whole multiplex network is estimated as the average of the entropies of the individual layers:

$$h_M(C) = \sum_{\alpha=1}^{|C|} h_\alpha / |C| \qquad (7.17)$$

where the configuration index C labels a specific combination of $|C|$ different layers, either original or aggregated. Because of the definition of entropy, the ratio $h_M(C)/h_A$ is upper-bounded by 1 ($h_M(C) = h_A$ when all the layers have been aggregated together). Therefore, the rescaled index,

$$q(C) = 1 - \frac{h_M(C)}{h_A} \qquad (7.18)$$

represents a relative entropy ranging between 0 and a given maximum value. $q(C) = 0$ when the aggregate network is equivalent to the multiplex in configuration C and

---

[1] Here, D is a diagonal matrix having node degrees on its main diagonal, K is the number of links in the network and A is the adjacency matrix.

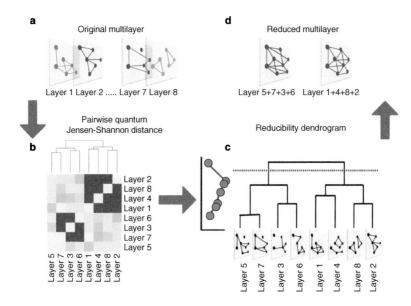

Figure 7.5   Example of structural reducibility analysis of a multi-layer network of $L = 8$ layers. Individual layers (a) are initially clustered according to how similar their Von Neumann entropies are (b); see also the subsection 'Layer Clustering'. A greedy algorithm selects the configurations in which the most similar layers are clustered first and it produces a hierarchy of configurations (c). The profile of the relative entropy q is maximized when only two aggregated layers are considered, thus suggesting that the multi-layer structure can be condensed into two layers only (d). Figure from De Domenico et al. [6]. © 2015, Springer Nature.

this can happen if and only if $|C| = 1$ and all layers have been aggregated (unless the multiplex network contains identical layers). $q(\cdot)$ is maximum when the multiplex structure in configuration C is maximally different or the most distinguishable from the aggregated network.

$q(\cdot)$ is an entropy-based approach for defining how different a given multiplex configuration is compared to its aggregated counterpart. In general, $q(\cdot)$ can either increase or decrease when two layers are aggregated, depending on their connectivity patterns. $q(\cdot)$ decreases when $h_M(C)$ increases and this happens mainly in two cases: (1) when two layers with very different link densities are aggregated into one layer; or (2) when new structural patterns that were absent in the original layers emerge in the aggregated one. Both of these cases imply a loss of topological information; this is why configurations that minimize h_M and thus maximize the relative entropy $q(\cdot)$ are preferred. This criterion identifies the multiplex configuration providing the most information about the multi-relational patterns within it. Figure 7.5 reports an example of structural reducibility analysis.

When $q(\cdot)$ is maximum but no aggregation is performed, that is layers are kept separate, then the original multiplex configuration is considered as *irreducible*: performing any layer aggregation would imply a loss of structural information. This is an important starting point for justifying the multiplex approach [17], together with the interpretation of what individual layers represent in the irreducible network representation.

Layer Clustering

One issue of structural reducibility analysis is that the number of different configurations that have to be tested to exhaustively discover the optimal configuration is the Lth Bell number for a multiplex with L layers, which increases super-exponentially with L. Testing a multiplex network with only $L = 15$ layers would require testing more than 10^9 configurations.

Layer clustering is a viable heuristics for solving the above issue [6]. Structural reducibility adopts a greedy algorithm exploring the configurations where more similar layers are aggregated first. Although there are different measures of network similarity in the literature (we refer the interested reader to [48] for a review), structural reducibility employs a similarity metric based on the Von Neumann entropy h which is commonly used in quantum mechanics, namely the Jensen–Shannon divergence:

$$\mathcal{D}_{JS}(\rho, \sigma) = h(\frac{\rho + \sigma}{2}) - \frac{h(\rho) + h(\sigma)}{2} \tag{7.19}$$

where ρ and σ are matrices. It is easy to check that $\sqrt{\mathcal{D}_{JS}}$ is bounded within $[0, 1]$ and satisfies the definition of a metric. When using the degree-normalized Laplacian matrices of two distinct networks, the Jensen–Shannon distance $\sqrt{\mathcal{D}_{JS}}$ can be used to implement a greedy algorithm performing structural reducibility among the most similar layers.

More in general, the Jensen–Shannon distance can be used also to perform hierarchical clustering of layers in multiplex networks, potentially unravelling community-based classifications across layers of network interactions [7]. Distances between empirical network layers and families of null models can also be successfully used for quantifying how much information is needed for correctly learning the parameters of a model [7].

7.5.5   Software for Multi-layer Network Structure Analysis

Over the years, the network community has developed several libraries and software for the analysis of the multi-layer network features discussed in the previous sections. We list these tools here for the interested reader:

- MuxViz (https://github.com/manlius/muxViz) is a self-contained framework for the structural and dynamical analysis and visualization of multi-layer networks, based on R [49].
- PymNet (www.mkivela.com/2015/12/11/multilayer-networks-library/) is a Python library for multi-layer network analysis integrated with NetworkX for the analysis of single-layer networks.
- MAMMULT (https://github.com/KatolaZ/mammult) is a collection of libraries in C and Python focusing on multiplex networks.

- GenLouvain (http://netwiki.amath.unc.edu/GenLouvain/GenLouvain) is a library in MATLAB implementing multislice community analysis.
- InfoMAP (www.mapequation.org/code.html) is a command line software implemented in C++ for multi-layer network analysis based on random walks.
- MultiNet (http://multilayer.it.uu.se/software.html) is a set of libraries implemented in R for multi-layer network analysis.

References

[1] De Domenico M, Solé-Ribalta A, Cozzo E, et al. Mathematical formulation of multilayer networks. *Phys Rev X*. 2013;3:041022.

[2] Cardillo A, Gómez-Gardenes J, Zanin M, et al. Emergence of network features from multiplexity. *Sci Rep*. 2013;3.

[3] Battiston F, Nicosia V, Latora V. Structural measures for multiplex networks. *Phys Rev E*. 2014;89(3):032804.

[4] Nicosia V, Latora V. Measuring and modeling correlations in multiplex networks. *Phys Rev E*. 2015;92(3):032805.

[5] Gemmetto V, Garlaschelli D. Multiplexity versus correlation: the role of local constraints in real multiplexes. *Sci Rep*. 2015;5.

[6] De Domenico M, Nicosia V, Arenas A, Latora V. Structural reducibility of multilayer networks. *Nat Commun*. 2015;6:6864.

[7] De Domenico M, Biamonte J. Spectral entropies as information-theoretic tools for complex network comparison. *Phys Rev X*. 2016;6:041062.

[8] Mucha PJ, Richardson T, Macon K, Porter MA, Onnela J-P. Community structure in time-dependent, multiscale, and multiplex networks. *Science*. 2010;328(5980):876–878.

[9] De Domenico M, Lancichinetti A, Arenas A, Rosvall M. Identifying modular flows on multilayer networks reveals highly overlapping organization in interconnected systems. *Phys Rev X*. 2015;5(1):011027.

[10] Kivelä M, Arenas A, Barthelemy M, et al. Multilayer networks. *J Complex Netw*. 2014;2(3):203–271.

[11] Boccaletti S, Bianconi G, Criado R, et al. The structure and dynamics of multilayer networks. *Phys Rep*. 2014;544(1):1–122.

[12] Barabási A-L. *Network Science*. Cambridge: Cambridge University Press, 2015.

[13] De Domenico M, Solé-Ribalta A, Gómez S, Arenas A. Navigability of interconnected networks under random failures. *PNAS*. 2014;111(23):8351–8356.

[14] Buldyrev SV, Parshani R, Paul G, Stanley HE, Havlin S. Catastrophic cascade of failures in interdependent networks. *Nature*. 2010;464(7291):1025–1028.

[15] Bianconi G, Dorogovtsev SN, Mendes JFF. Mutually connected component of networks of networks with replica nodes. *Phys Rev E.* 2015;91(1):012804.

[16] Baxter GJ, Cellai D, Dorogovtsev SN, Goltsev AV, Mendes JFF. A unified approach to percolation processes on multiplex networks. In: *Interconnected Networks.* New York: Springer, 2016.

[17] Stella M, Beckage NM, Brede M, De Domenico M. Multiplex model of mental lexicon reveals explosive learning in humans. *Sci Rep.* 2018;8(1):2259.

[18] Baggio JA, BurnSilver SB, Arenas A, et al. Multiplex social ecological network analysis reveals how social changes affect community robustness more than resource depletion. *PNAS.* 2016;113(48):13708–13713.

[19] Reis SDS, Hu Y, Babino A, et al. Avoiding catastrophic failure in correlated networks of networks. *Nat Phys.* 2014;10(10):762–767.

[20] De Domenico M, Granell C, Porter MA, Arenas A. The physics of spreading processed in multilayer networks. *Nat Phys.* 2016;12:901–906.

[21] Stella M, Andreazzi CS, Selakovic S, Goudarzi A, Antonioni A. Parasite spreading in spatial ecological multiplex networks. *J Complex Netw.* 2016;5(3):486–511.

[22] Scott J. *Social Network Analysis.* Thousand Oaks, CA: Sage, 2017.

[23] Solá L, Romance M, Criado R, et al. Eigenvector centrality of nodes in multiplex networks. *Chaos.* 2013;23(3):033131.

[24] Stella M, Beckage NM, Brede M. Multiplex lexical networks reveal patterns in early word acquisition in children. *Sci Rep.* 2017;7.

[25] Battiston F, Nicosia V, Latora V. The new challenges of multiplex networks: measures and models. *Eur Phys J Special Top.* 2017;226(3):401–416.

[26] Bianconi G. Statistical mechanics of multiplex networks: entropy and overlap. *Phys Rev E.* 2013;87(6):062806.

[27] Guimera R, Nunes Amaral LA. Functional cartography of complex metabolic networks. *Nature.* 2005;433(7028):895–900.

[28] Stella M, Selakovic S, Antonioni A, Andreazzi CS. Community interactions determine role of species in parasite spread amplification: the ecomultiplex network model. *arXiv preprint arXiv:1706.05121,* 2017.

[29] De Domenico M, Solé-Ribalta A, Omodei E, Gómez S, Arenas A. Ranking in interconnected multilayer networks reveals versatile nodes. *Nat Commun.* 2015;6:6868.

[30] Halu A, Mondragón RJ, Panzarasa P, Bianconi G. Multiplex pagerank. *PLoS One.* 2013;8(10):e78293.

[31] Katz L. A new status index derived from sociometric analysis. *Psychometrika.* 1953;18(1):39–43.

[32] Kleinberg JM. Authoritative sources in a hyperlinked environment. *J ACM.* 1999;46(5):604–632.

[33] Page L, Brin S, Motwani R, Winograd T. The PageRank citation ranking: bringing order to the web. Technical report, Stanford InfoLab, 1999.

[34] Iacovacci J, Rahmede C, Arenas A, Bianconi G. Functional multiplex PageRank. *EPL*. 2016;116(2):28004.

[35] Battiston F, Nicosia V, Latora V. Efficient exploration of multiplex networks. *New J Phys*. 2016;18(4):043035.

[36] Omodei E, De Domenico M, Arenas A. Evaluating the impact of interdisciplinary research: a multilayer network approach. *Netw Sci*. 2017;5(2):235–246.

[37] Battiston F, Iacovacci J, Nicosia V, Bianconi G, Latora V. Emergence of multiplex communities in collaboration networks. *PLoS One*. 2016;11(1):e0147451.

[38] Cozzo E, Kivelä M, De Domenico M, et al. Structure of triadic relations in multiplex networks. *New J Phys*. 2015; 17(7):073029.

[39] Baxter GJ, Cellai D, Dorogovtsev SN, Mendes JFF. Cycles and clustering in multiplex networks. *Phys Rev E*. 2016;94(6):062308.

[40] Fortunato S. Community detection in graphs. *Phys Rep*. 2010;486(3):75–174.

[41] Newman MEJ. Modularity and community structure in networks. *PNAS*. 2006;103(23):8577–8582.

[42] Rosvall M, Bergstrom CT. Maps of random walks on complex networks reveal community structure. *PNAS*. 2008;105(4):1118–1123.

[43] Von Luxburg U. A tutorial on spectral clustering. *Stat Comput*. 2007;17(4): 395–416.

[44] De Domenico M. Diffusion geometry unravels the emergence of functional clusters in collective phenomena. *Phys Rev Lett*. 2017;118(16):168301.

[45] Gauvin L, Panisson A, Cattuto C. Detecting the community structure and activity patterns of temporal networks: a non-negative tensor factorization approach. *PLoS One*. 2014;9(1):e86028.

[46] Radicchi F, Bianconi G. Redundant interdependencies boost the robustness of multiplex networks. *Phys Rev X*. 2017;7(1):011013.

[47] Passerini F, Severini S. Quantifying complexity in networks: the Von Neumann entropy. *Int J Agent Tech Syst*. 2009;1(4):58–67.

[48] Bródka P, Chmiel A, Magnani M, Ragozini G. Quantifying layer similarity in multiplex networks: a systematic study. *arXiv preprint arXiv:1711.11335*, 2017.

[49] De Domenico M, Porter MA, Arenas A. Muxviz: a tool for multilayer analysis and visualization of networks. *J Complex Netw*. 2015;3(2):159–176.

8 Dynamics on Multi-layer Networks

Manlio De Domenico and Massimo Stella

8.1 Introduction

The multi-layer organization of complex systems (Figure 8.1) leads, in general, to a richer dynamics than single-layer/monoplex networks, because it allows us either to describe a single dynamical process or to couple different ones, on top of such interconnected structures. The interplay between structure and dynamics in multi-layer networks, as well as the interplay between distinct dynamical processes, has revealed a new set of physical phenomena that have no counterpart in monoplex networks.

In this chapter, we review the most recent advances in modeling and analysis of multi-layer network dynamics. However, we will not cover the results on the dynamics of percolation [1–9], complex spreading models [10–15], evolutionary game theory [16–19], consensus [20, 21], synchronization [22–26], reaction-diffusion [27, 28] and growing models [29, 30], which would deserve dedicated chapters. We refer the interested reader to thorough reviews in [31–36].

Single vs coupled dynamics. Dynamics on the top of multi-layer networks can be classified into two main categories: (1) a single dynamical process within and across layers (Figure 8.2a); and (2) "mixed" or "coupled" dynamics, where a different dynamical process is defined on each layer separately and all processes are intertwined by inter-layer connectivity (Figure 8.2b).

In the case of molecular networks, single-layer dynamics appears to be more appropriate for modeling purposes and applications. Here, we briefly review existing coupled dynamics for the sake of completeness, and we refer the interested reader to the corresponding literature. Coupled processes investigated so far are mostly related to spreading dynamics and play a fundamental role in understanding phenomena of concurrent diseases in multiplex networks [12, 37–39], and of information or behavior spreading in interdependent or multiplex sociotechnical systems [14, 40, 40–44]. This type of dynamics is interesting because they can enhance or inhibit each

(a)   (b)   (c)

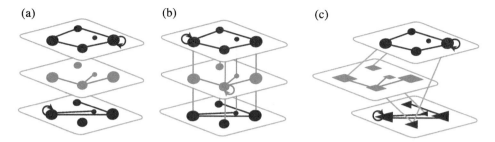

Figure 8.1 Multi-layer network topologies. Different multi-layer network topologies might be characterized by different dynamical properties. (a) Edge-colored network, characterized by layers encoding different relationships (i.e. colors) and the absence of inter-layer connectivity. Here, information among nodes' replicas is assumed to be transmitted instantaneously. (b) Multiplex network, an edge-colored topology in which inter-layer connectivity between nodes and their replicas on other layers is present. Information among nodes' replicas is transmitted with a cost, encoded by the weight of inter-layer links. (c) Interdependent network, characterized by nodes of different types (circles, squares and triangles) and inter-layer connectivity without additional topological constraints. Figure from [35]. Springer Nature, 2016.

(a)   (b)

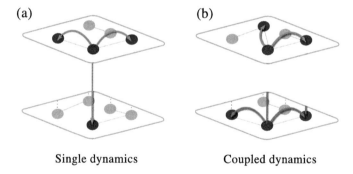

Single dynamics   Coupled dynamics

Figure 8.2 Dynamical processes on multi-layer networks. (a) A single dynamical process (encoded in arcs' colors) running within and across layers of a multi-layer network. (b) Two different dynamical processes (encoded by different arcs' colors, one per layer), running on different layers. The processes are coupling (or intertwined) by the inter-layer connectivity. Figure from [35]. Springer Nature, 2016.

other [14, 38, 43], leading to new physical phenomena such as the existence of a curve of critical points that separate different phases – where the critical properties of one process depend or not on those of the other [38, 43] – to the emergence of consensus subject to social or cognitive constraints [21] and other peculiar collective phenomena [45].

In the following, we mostly focus on dynamical processes of interest for modeling molecular networks and biological integrated systems [46], namely diffusion processes, and we discuss their application for (1) identifying essential units for information flow; (2) defining the navigability of multi-layer topologies; and (3) unraveling the mesoscale structure.

8.2   Single Dynamics on Multi-layer Networks

In single dynamics, we analyze physical phenomena that arise from a single dynamical process on top of a multi-layer structure. The behavior of such a process depends both on intra-layer structure (i.e. the usual considerations in networks) and on inter-layer structure (i.e. the presence and strength of interactions between nodes on different layers).

In this section, we discuss physical phenomena arising from two different single dynamical process, continuous and discrete diffusion, representative for modeling a wide variety of scenarios. We can anticipate that the overall behavior of such processes depends, in general, on the interplay between intra- and inter-layer connectivity. In fact, depending on the relative importances of both types of connection, a multi-layer system can act either as a set of structurally decoupled layers or as a single-layer system, in which layers are indiscernible in practice. Under specific conditions, a sharp transition between these two regimes [47, 48] has been observed.

8.2.1   Continuous-Time Non-normalized Laplacian Dynamics: Diffusion

In the following we will distinguish between physical nodes, corresponding to a specific entity within the network regardless of its replicas across layers, and state nodes, corresponding to entities' replicas across layers. Vectors encoding information about physical and state nodes are denoted by x_i ($i = 1, 2, \ldots, N$, with N being the size of the network) and $x_{i\alpha}$ ($\alpha = 1, 2, \ldots, M$, with M being the number of layers). It is worth remarking that in both cases we are dealing with vectors and that, for the sake of simplicity, we make an abuse of notation by indicating with the same symbol the vector's components.

Let us start from the diffusion equation on the top of a single-layer/monoplex network. We denote the state vector of nodes by $x_i(t)$, encoding the information present in each node at time t. Assuming an initial state $x_i(0)$ and Einstein notation, the evolution of the state vector according to the simplest diffusive continuous dynamics is governed by the diffusion equation

$$\frac{dx_j(t)}{dt} = D\left[W_j^i x_i(t) - W_k^i u_i e^k(j) x_j(t)\right],\tag{8.1}$$

where D is a diffusion constant, u_i is the vector of 1s and $e^k(j)$ is the canonical vector with components equal to 0, except for the jth one. By recalling that the vector encoding the strength of each node is defined by $s_k = W_k^i u_i$ and that $s_k e^k(j) x_j(t) = s_k e^k(j) \delta_j^i x_i(t)$, the diffusion equation can be written as:

$$\frac{dx_j(t)}{dt} = -DL_j^i x_i(t),\tag{8.2}$$

where $L_j^i = W_k^l u_l e^k(j) \delta_j^i - W_j^i$ is the combinatorial Laplacian tensor [49]. The solution of Eq. (8.2) is given by $x_j(t) = x_i(0) e^{-DL_j^i t}$. This short introduction allows one to exploit the tensorial formalism [50] to build the diffusion equation for a multi-layer network.

A first pioneering study about diffusion in interconnected multiplex networks, based on the concept of supra-adjacency matrix, was reported by Gomez et al. [51]. Here, we make use of the tensorial formulation, generalizing the study of this dynamical process to any multi-layer system [50].

At variance with the monoplex case, information in multi-layer networks also diffuses through inter-layer links. If we indicate by $X_{i\alpha}(t)$ the state tensor of replica nodes in each layer at time t, the diffusion equation for a multi-layer network is then

$$\frac{dX_{j\beta}(t)}{dt} = M_{j\beta}^{i\alpha} X_{i\alpha}(t) - M_{k\gamma}^{i\alpha} U_{i\alpha} E^{k\gamma}(i\beta) X_{i\beta}(t), \tag{8.3}$$

where $U_{i\alpha} = u_i u_\alpha$ is the tensor of 1s and $E^{k\gamma}(i\beta) = e^k(i)e^\gamma(\beta)$ is the canonical tensor equivalent to a matrix with 0 entries, except for the ith row and βth column. By defining the multi-layer combinatorial Laplacian as

$$L_{j\beta}^{i\alpha} = M_{k\gamma}^{l\epsilon} U_{l\epsilon} E^{k\gamma}(j\beta)\delta_{j\beta}^{i\alpha} - M_{j\beta}^{i\alpha}, \tag{8.4}$$

we can write the covariant diffusion equation for multi-layer networks more compactly as

$$\frac{dX_{j\beta}(t)}{dt} = -L_{j\beta}^{i\alpha} X_{i\alpha}(t), \tag{8.5}$$

whose solution is given by $X_{j\beta}(t) = X_{i\alpha}(0)e^{-L_{j\beta}^{i\alpha}t}$, providing a natural generalization of the result for monoplex networks. From the analysis of the spectral properties of the Laplacian tensor, it can be shown that the speed of diffusion is governed by its second smallest eigenvalue Λ_2 [22, 50, 51], calculated from the supra-adjacency representation of the multi-layer network obtained by flattening [52] the rank-4 tensor. This feature is very important to gain new insights from the comparison between diffusion properties in the multi-layer system and in its aggregation to a single-layer network.

In fact, it has been shown [51] that in the case of a two-layer multiplex network, it is possible to observe a faster diffusion than in each layer separately, if the condition

$$\Lambda_2^{\text{multiplex}} \geq \max\{\Lambda_2^{\text{layer 1}}, \Lambda_2^{\text{layer 2}}\} \tag{8.6}$$

is satisfied. Here, we show an application to a synthetic system consisting of two Erdős–Rényi (ER) networks, characterized by wiring probability p_1 and p_2, respectively (Figure 8.3). As shown in the central panel of the figure, this condition holds when $p_1 \approx p_2$, that is when the two layers are characterized by a similar amount of homogeneity. To better understand the role of inter-layer weight, connections between nodes and their replicas are weighted by ω, ranging between 0.1 and 110 in this example. The left (right) panel of Figure 8.3 shows the behavior of Λ_2 as a function of the inter-layer coupling weight ω when $p_1 \approx p_2$ ($p_1 \neq p_2$). In both cases it is possible to observe a sharp change in the value of the Λ_2, separating two different regimes that correspond to different structural properties of the multi-layer network. However, in one case (left panel) multi-layer diffusion is faster than diffusion in each layer separately above a certain critical value of ω, whereas in the other case (right panel) it is slower, because $\Lambda_2^{\text{layer 2}} > \Lambda_2^{\text{multiplex}}$ regardless of the value of ω. The transition between the two regimes is a structural transition [47], a characteristic of multi-layer networks also observed in other contexts [53].

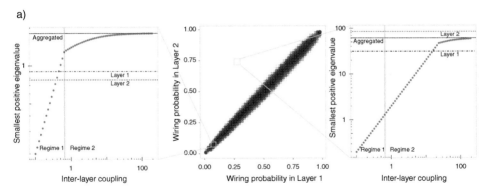

Figure 8.3   Diffusion dynamics on a multi-layer network. The second smallest eigenvalue of a combinatorial Laplacian tensor, generally indicated by Λ_2, governs the speed of continuous-time Laplacian diffusion dynamics in multi-layer networks. In this example, two ER networks – characterized by independently varying probabilities $p_1, p_2 \in [0, 1]$ to connect two nodes within the same layer – are coupled together in a multiplex network. Depending on the values of the two parameters, the diffusion speed in the multi-layer system can be faster (left) or slower (right) than in each layer, separately. Figure from [35]. Springer Nature, 2016.

8.2.2   Discrete-Time Normalized Laplacian Dynamics: Random Walks

Diffusion can also occur in discrete processes. A canonical example is given by random walks [55–57], modeling Markovian dynamics on top of the network (refer to [58] for a thorough review). The importance of random walks is due to their analytical tractability that, in many applications, can be used to approximate more complex dynamics.

In classical networks, a walker can only jump from one node to another in its close neighborhood, although there are walks allowing for direct teleportation to any nodes [59, 60], even those in other connected components of the system, as we will see later. In multi-layer networks, a walker *jumps* between nodes within the same layer through intra-layer edges and can also *switch* between nodes on different layers through inter-layer edges to explore the whole interconnected system [48, 54, 61] (Figure 8.4).

In a random walk, the walker is initially located in a certain node and moves to other nodes following specific transition rules encoded into a matrix named *transition matrix*. In the simplest scenario [56, 62], the rule is to choose uniformly randomly among edges outgoing from the node where the walker is positioned, but more complex rules are possible, as we will see in the following. From a mathematical point of view, the dynamics of a random walk is governed by a master equation, which has been recently generalized to the case of multi-layer networks [50, 54].

As for the case of Laplacian dynamics, there is a tensor governing the dynamics, which we indicate by $\mathcal{P}^{i\alpha}_{j\beta}$, which encodes the probability for jumping/switching from node i in layer α to node j in layer β ($i, j = 1, 2, \ldots, N$ are Latin indices to indicate nodes and $\alpha, \beta = 1, 2, \ldots, M$ are Greek indices to indicate layers, as in the previous section). Let $p_{i\alpha}(t)$ denote the probability of finding the random walker in node i of

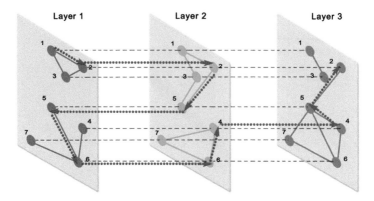

Figure 8.4   Random walk on an interconnected multiplex network. The dotted red line encodes the path of a random walker exploring the network. In this example, a walk through disconnected components in a single layer is allowed by the multiplex structure. In this example, the walker is not allowed to switch between layer 1 and layer 3 in one time step. Figure from De Domenico et al. [54].

layer α at time t. Making use again of Einstein summation convention, the master equation reads

$$p_{j\beta}(t + \Delta t) = \mathcal{P}^{i\alpha}_{j\beta} p_{i\alpha}(t). \tag{8.7}$$

It is worth separating the contribution of jumping and switching movements in the random walk dynamics:

$$p_{j\beta}(t + \Delta t) = \underbrace{\mathcal{P}^{j\beta}_{j\beta} p_{j\beta}(t)}_{\text{stay in the node}} + \underbrace{\sum_{\substack{\alpha=1 \\ \alpha \neq \beta}}^{L} \mathcal{P}^{j\alpha}_{j\beta} p_{j\alpha}(t)}_{\text{switch to replica node}} + \underbrace{\sum_{\substack{i=1 \\ i \neq j}}^{N} \mathcal{P}^{i\beta}_{j\beta} p_{i\beta}(t)}_{\text{jump to neighbor}} + \underbrace{\sum_{\substack{\alpha=1 \\ \alpha \neq \beta}}^{L} \sum_{\substack{i=1 \\ i \neq j}}^{N} \mathcal{P}^{i\alpha}_{j\beta} p_{i\alpha}(t)}_{\text{switch and jump}}.$$

It is possible to show that Eq. (8.7) leads to the following differential equation:

$$\frac{dp_{j\beta}(t)}{dt} = -\tilde{L}^{i\alpha}_{j\beta} p_{i\alpha}(t), \tag{8.8}$$

where $\tilde{L}^{i\alpha}_{j\beta} = \mathcal{I}^{i\alpha}_{j\beta} - \mathcal{P}^{i\alpha}_{j\beta}$ is known as the normalized Laplacian tensor, with $\mathcal{I}^{i\alpha}_{j\beta}$ being the identity tensor. The reader should notice the similarity with the master equation in the case of continuous diffusion, the most relevant difference being about the tensor governing the underlying dynamics.

The dynamics of any random walk, regardless for the peculiar transition rules that characterize it, can be described by the master equation in Eq. (8.8). In Section 8.3 we will analyze different transition rules that encode very different random walk dynamics, used to model information flow in multi-layer networks.

Node Occupation Probability

For multi-layer networks characterized by undirected relationships and consisting of a single multi-layer connected component, all nodes have a certain probability of being visited. For instance, in the case of a classical random walk on the top of an undirected and unweighted network, where edges are followed uniformly randomly, the probability of a state node i being visited in layer α is proportional to its degree $k_{i\alpha}$ in the stationary state; that is, after an infinite amount of time.

The occupation probability of node i in layer α is defined as the probability of finding the random walker in that location of the multiplex, in the limit of large time, $\Pi_{i\alpha} = \lim_{t\to\infty} p_{i\alpha}(t)$. We also indicate with Π the corresponding supra-vector. In general, Π is the left eigenvector of the supra-transition matrix corresponding to the unit eigenvalue. A traditional approach to calculate the occupation probability is based on the detailed balance equation:

$$\Pi_{i\alpha} P_{j\beta}^{i\alpha} = \Pi_{j\beta} P_{i\alpha}^{j\beta}. \tag{8.9}$$

It is worth remarking that indices in Eq. (8.9) indicate the corresponding entries and Einstein summation convention is not applied: the equation is not tensorial. Once the transition rules are established (i.e. the tensor $P_{j\beta}^{i\alpha}$ is fixed), it is possible to use the detailed balance equation to estimate $\Pi_{i\alpha}$, encoding the occupation probability of state node i in layer α. The occupation probability for the underlying physical node i is simply given by $\pi_i = \Pi_{i\alpha} u^{\alpha}$; that is, by summing up the probabilities corresponding to each state node [63], under the assumption that all layers have the same importance [64, 65].[1] In Section 8.3, we report the analytical calculations for some representative random walk dynamics.

Mean Return Time and Mean First Passage Time

In the same spirit as the approach proposed in [56] for random walks on single-layer networks, it is straightforward to show that the time required for a random walker originating from node i to arrive back at the same node regardless of the layer (i.e. the mean return time) is given by

$$\langle T_{ii} \rangle = \frac{1}{\Pi_{i\alpha} u^{\alpha}}. \tag{8.10}$$

The calculation of mean first passage time (MFPT), defined as the average number of steps to reach a node d, starting from a given node s, requires more effort [65]. In monoplex networks, the matrix encoding MFPT for all pairs of nodes in the systems can be computed analytically by means of Kemeny–Snell fundamental matrix Z [66, 67] or by means of absorbing random walks [68]. In the following, we will briefly discuss the second approach, involving the transition tensor P governing random walks over multi-layer networks and the corresponding absorbing transition tensor $P_{[d]}$ on the physical node d. The tensor

$$p_{j\beta}^{o\sigma}(t) = \left(P_{[d]}^{t} \right)_{j\beta}^{o\sigma} \tag{8.11}$$

encodes the probability that the walker visits node j in layer β, after t time steps, assuming that it originated in node o in layer σ. It is worth remarking that $p_{d\beta}^{o\sigma}(t) = 0$ for any β and t, because the transition tensor is absorbing on node d. The probability

---

[1] If this is not the case, a weighted average can be used instead.

that the walker is absorbed in some physical node d at a time T equal to or smaller than t is given by

$$\left(q_{[d]}\right)^{o\sigma}(t) = u^{o\sigma} - \left(P_{[d]}^{t}\right)_{j\beta}^{o\sigma} u^{j\beta}. \tag{8.12}$$

This equation defines a rank-2 tensor q for each choice of node d and, for each tensor, the probability that the first passage time for node d is exactly t is given by

$$\left(q_{[d]}\right)^{o\sigma}(T = t) = \left(q_{[d]}\right)^{o\sigma}(t) - \left(q_{[d]}\right)^{o\sigma}(t-1) = \left[\left(P_{[d]}^{t}\right) - \left(P_{[d]}^{t-1}\right)\right]_{j\beta}^{o\sigma} u^{j\beta}. \tag{8.13}$$

By assuming that the walk originates in node o and layer σ, each tensor encoding the MFPT to node d is obtained from Eq. (8.13) as

$$\left(m_{[d]}\right)^{o\sigma} = \sum_{t=0}^{\infty} t \left(q_{[d]}\right)^{o\sigma}(T = t) = \left[(\delta - P_{[d]}^{t})^{-1}\right]_{j\beta}^{o\sigma} u^{j\beta}. \tag{8.14}$$

Since the maximum eigenvalue of $P_{[d]}$ is strictly smaller than 1, the geometric series in Eq. (8.14) must converge to a finite value. It is worth noting here that the MFPT to node d still depends on the initial conditions; that is, node o and layer σ. To calculate the expected MFPT $H_{[d]}$ to node d, regardless of where the walk originated, it is sufficient to average $\left(m_{[d]}\right)^{o\sigma}$ over all possible starting nodes and layers as

$$H_{[d]} = \frac{1}{(N-1)M} u_{o\sigma} \left(m_{[d]}\right)^{o\sigma} + \frac{1}{N}\pi_d^{-1}, \tag{8.15}$$

where π_d is the occupation probability of node d and the term $\frac{1}{N}\pi_d^{-1}$ accounts for the average return time that is not accounted for when using absorbing random walks.

8.3   Modeling Information Flow in Multi-layer Networks

Information flow in multi-layer networks can be used for a variety of applications [32, 33, 35]. In most cases, even complex dynamics can be approximated, at first order, by Laplacian dynamics.

In the following, let us consider a multi-layer system with N nodes in each of the M layers; we use $W_{ij}^{(\alpha)}$ to indicate the weighted intra-layer connection between two vertices i and j in layer α, where Latin letters refer to vertices ($i, j = 1, 2, \ldots, N$) and Greek letters indicate layers ($\alpha = 1, 2, \ldots, M$). Note that the requirement that all nodes must be present in all layers can be easily relaxed [50]. Let $D_{(i)}^{\alpha\beta}$ denote the weight of switching from layer α to layer β when the walker is positioned in node i. Without loss of generality, we may suppose that $W_{ii}^{(\alpha)} = 0$ for all nodes i, since these self-loops can be accounted for in the terms $D_{(i)}^{\alpha\alpha}$. Note that the above matrices are not multi-layer adjacency tensors: at this step, we make use of the supra-adjacency formalism [22, 51] to highlight the role played by intra- and inter-layer edges in the walk dynamics.

The supra-adjacency is a (possibly weighted) $NM \times NM$ matrix defined in block format by

$$
\mathcal{A} = \begin{pmatrix}
\mathbf{D}^{11} + \mathbf{W}^{(1)} & \mathbf{D}^{12} & \cdots & \mathbf{D}^{1M} \\
\mathbf{D}^{21} & \mathbf{D}^{22} + \mathbf{W}^{(2)} & \cdots & \mathbf{D}^{2M} \\
\vdots & \vdots & \ddots & \vdots \\
\mathbf{D}^{M1} & \mathbf{D}^{M2} & \cdots & \mathbf{D}^{MM} + \mathbf{W}^{(M)}
\end{pmatrix},
\tag{8.16}
$$

where $\mathbf{W}^{(\alpha)}$ is the matrix encoding intra-layer weights of layer α and $\mathbf{D}^{\alpha\beta}$ is a diagonal matrix of inter-layer weights, encoding the cost for switching from layer α to layer β. In the simplest scenario, common to a wide variety of real systems, inter-layer weights are equal regardless of the node – that is, $\mathbf{D}^{\alpha\beta} = D^{\alpha\beta}\mathbf{I}$, where $\mathbf{I}$ is the $N \times N$ identity matrix. Very often, $D^{\alpha\beta} = 1$.

In order to build some transition matrices, it is useful to introduce at this step the ith node's intra-layer strength in layer α by $s_{i\alpha} = \sum_j W_{ij}^{(\alpha)}$, and its inter-layer strength in layer α by $S_{i\alpha} = \sum_\beta D_{(i)}^{\alpha\beta}$, assuming the general case of a weighted network.

8.3.1 Classical Random Walks

The classical description of random walkers on a graph (i.e. monoplex networks) is presented in [66, 69, 70], although applications to networks with complex topologies are more recent [56, 62].

The direct extension of such walks to the case of multiplex networks is obtained by considering the inter-layer connections as additional edges that can be used by the random walker once positioned in node i. In agreement with classical random walks, we impose that the probability of moving from node i to node j within the same layer α or to switch to another state node i in layer β is uniformly distributed. The entries of the multi-layer transition tensors for classical random walks (RWC) are reported in Table 8.1.

It is straightforward to calculate that node's occupation probability for RWC is given by

$$
\Pi_{i\alpha} = \frac{s_{i\alpha} + S_{i\alpha}}{\sum_\beta \sum_j (s_{j\beta} + S_{j\beta})},
\tag{8.17}
$$

generalizing the well-known result obtained for walks in single-layer networks.

8.3.2 Diffusive Random Walks

In single-layer networks, diffusive random walks (RWD) have been analyzed in [71]. In this case, the random walker stays in node i with a rate that depends on the node's strength. By defining the maximum strength in the system as $s_{\max} = \max_{i,\alpha}\{s_{i\alpha} + S_{i\alpha}\}$, the walker waits in node i on layer α with rate $1 - \tilde{s}_{i\alpha}/s_{\max}$, with $\tilde{s}_{i\alpha} = D_{(i)}^{\alpha\alpha} - s_{i\alpha} - S_{i\alpha}$ and jumps to any neighbor with rate proportional to $1/s_{\max}$ (see Table 8.1 for details). It can be shown that this dynamics is equivalent to one on the top of a network where self-loops are added to any node in such a way that all nodes have the same strength.

Table 8.1 Transition probability for four different random walk processes on multiplex networks. We account for jumping between vertices (Latin letters) and switching between layers (Greek letters). When appearing in pairs, $j \neq i$ and $\beta \neq \alpha$ must be considered. See text for further details.

Tr.	RWC	RWD	RWP	RWME
$\mathcal{P}^{i\alpha}_{i\alpha}$	$\dfrac{D^{\alpha\alpha}_{(i)}}{s_{i\alpha} + S_{i\alpha}}$	$\dfrac{s_{max} + D^{\alpha\alpha}_{(i)} - s_{i\alpha} - S_{i\alpha}}{s_{max}}$	0	$\dfrac{D^{\alpha\alpha}_{(i)}}{\lambda_{max}}$
$\mathcal{P}^{i\alpha}_{i\beta}$	$\dfrac{D^{\alpha\beta}_{(i)}}{s_{i\alpha} + S_{i\alpha}}$	$\dfrac{D^{\alpha\beta}_{(i)}}{s_{max}}$	0	$\dfrac{D^{\alpha\beta}_{(i)}}{\lambda_{max}} \dfrac{\psi_{(\beta-1)N+i}}{\psi_{(\alpha-1)N+i}}$
$\mathcal{P}^{i\alpha}_{j\alpha}$	$\dfrac{W^{(\alpha)}_{ij}}{s_{i\alpha} + S_{i\alpha}}$	$\dfrac{W^{(\alpha)}_{ij}}{s_{max}}$	$\dfrac{W^{(\alpha)}_{ij}}{s_{i\alpha}} \dfrac{D^{\alpha\alpha}_{(i)}}{S_{i\alpha}}$	$\dfrac{W^{(\alpha)}_{ij}}{\lambda_{max}} \dfrac{\psi_{(\alpha-1)N+j}}{\psi_{(\alpha-1)N+i}}$
$\mathcal{P}^{i\alpha}_{j\beta}$	0	0	$\dfrac{W^{(\beta)}_{ij}}{s_{i\beta}} \dfrac{D^{\alpha\beta}_{(i)}}{S_{i\alpha}}$	0

It is straightforward to show that RWD dynamics is governed by the combinatorial Laplacian, as in Laplacian diffusion: detailed balance calculations lead to the following node's occupation probability:

$$\Pi_{i\alpha} = \frac{1}{NM}. \tag{8.18}$$

That is, it is uniformly distributed and equal for all state nodes, as expected for a purely diffusive walk.

8.3.3 Physical Random Walks

A new type of random walk dynamics in the multiplex can be introduced, which has no counterpart on monoplex networks and reduces to the RWC when $M = 1$. The fundamental hypothesis is that the time scale required to switch layer is negligible with respect to the time scale required to jump onto a node in its neighborhood. In fact, the random walk can independently switch layer and jump to another node in the same time step. Another remarkable difference is that, at variance with RWC, inter-layer edges do not compete with intra-layer ones. Transition rules for this physical random walker (RWP) are reported in Table 8.1.

8.3.4 Maximum Entropy Random Walks

In classical random walks, a walker navigates the network according to rules based on the local structure: it does not require global knowledge of the system. Such a walk locally maximizes the entropy of paths, choosing uniformly randomly one edge among the available local ones.

However, it is possible to define transition rules based on global knowledge of the network [72], where the walker maximizes the entropy of global paths rather than local ones. More specifically, in this random walk dynamics all global paths are equally likely. To achieve such maximal entropy paths, the transition rates are governed by the largest eigenvalue of the adjacency matrix and the components of the corresponding eigenvector [72]. In the case of multi-layer networks, one can calculate the leading eigenvector of the supra-adjacency matrix for the same purpose, if we indicate with λ_{max} the largest eigenvalue of this matrix and with ψ the corresponding eigenvector. Transition rules for this maximum entropy random walk (RWME) are given in Table 8.1. Calculations from detailed balance allow one to calculate the node's occupation probability as follows:

$$\Pi_{i\alpha} = \psi^2_{(\alpha-1)N+i},$$

(8.19)

generalizing the results obtained for classical single-layer networks [72].

A representative example of each walk is shown in Figure 8.5 for 100 time steps. Two different cases are considered to highlight the influence of inter-layer weights on the dynamics.

A visual representation of the transition probabilities corresponding to the four random walks considered above is given in Figure 8.6. A multiplex network of 2 layers and 20 nodes, interconnected according to two different realizations of a Watts–Strogatz small-world network [73], is considered. The figure also shows the probability of finding the walker in any vertex, assuming that it originated from node 1 (middle panels) or from any other node with uniform probability (bottom panels). As expected, different exploration strategies result in different occupation probabilities, where some vertices in a certain layer might be explored more (or less) frequently, as in the case of RWC, RWP and RWME, or uniformly as in the case of RWD.

Other transition rules can be defined. For instance, it is possible to generalize biased random walks to edge-colored networks [74], systems characterized by the absence of inter-layer connectivity.

Figures 8.5 and 8.6 clearly highlight the different dynamics and how navigation strategy influences the exploration of multi-layer systems and, consequently, information diffusion.

8.3.5 Network Coverage

Random walks can be used to quantify how a network is difficult to explore. The measure used for this purpose is known as coverage $\rho(t)$, and it is defined by the average fraction of nodes visited at least once within a certain time t. Given the existence of multiple state nodes in multi-layer networks, a physical node is labeled as visited if at least one of its corresponding state nodes has been visited by the walker.

The probability of finding the random walker in node i at time t regardless of the layer is given by $p_i(t) = p_{i\alpha}(t)u^\alpha$. Indicating with $\mathbf{e}_i$ the ith canonical vector, let us define the supra-vector $\mathbf{E}_i \equiv (\mathbf{e}_i, \mathbf{e}_i, \ldots, \mathbf{e}_i)$ to rewrite the above probability as

$$p_i(t+1) = \mathbf{p}(t)\mathcal{P}\mathbf{E}_i^\dagger.$$

(8.20)

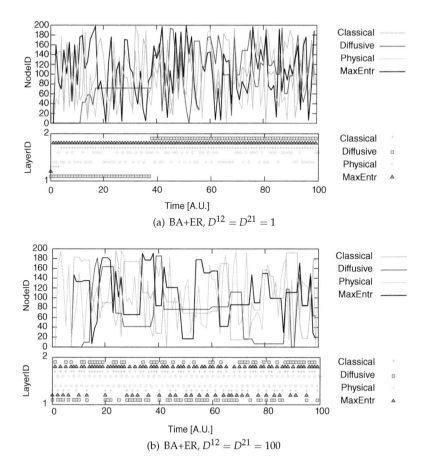

Figure 8.5   Random walks realizations on different multiplex structures with different inter-layer coupling. Nodes (top panels) and layers (bottom panels) visited by one random walker in 100 time steps. The four types of random walk dynamics considered in this section are shown. The multiplex is built with one Barabási-Albert (layer 1) and one ER (layer 2) network with 200 vertices, while inter-layer weights are equal to 1 (top) and 100 (bottom), respectively. Figure from De Domenico et al. [54].

Note that $\Delta t = 1$ to make a formal equivalence between time t and the walk length. Let $\sigma_{ij}(t)$ indicate the probability of not finding the random walker in node i after t time steps, assuming that it originated in node j. It is possible to show that the recursive relation

$$\sigma_{ij}(t+1) = \sigma_{ij}(t)\left[1 - p_i(t+1)\right] \tag{8.21}$$

holds and its solution is given by

$$\sigma_{ij}(t) = \sigma_{ij}(0)\exp\left[-\mathbf{p}_j(0)\mathbb{P}\mathbf{E}_i^{\dagger}\right], \quad \mathbb{P} = \sum_{\tau=0}^{t}\mathcal{P}^{\tau+1}, \tag{8.22}$$

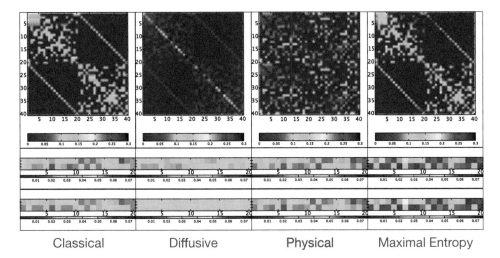

Figure 8.6 Transition rules for four multi-layer random walk dynamics. A multiplex network with 2 layers and 20 vertices, connected according to two different realizations of a Watts–Strogatz small-world network (rewiring probability is 0.2), where $D^{11} = D^{12} = D^{21} = D^{22} = 1$ is considered. *Top panels:* transition probability matrices. Note that we have rescaled by a factor 2 the transition matrix of diffusive walk for better visualization and to allow for comparisons. *Middle panels:* occupation probability, for each vertex in each layer, considering one random walk starting only from the first vertex. *Bottom panels:* as in the middle panels, but considering one random walk starting with uniform probability from any other vertex. Figure from De Domenico et al. [54].

where, without loss of generality, $\mathbf{p}_j(0) \equiv (\mathbf{e}_j, \mathbf{0}, \ldots, \mathbf{0})$ encodes the assumption that the walker started in node j and in the first layer at time $t = 0$. The matrix $\mathbb{P}$ encodes the probability to reach each node through any path of length $1, 2, \ldots, t + 1$. It might be advantageous to note that $\sigma_{ij}(0) = 1 - \delta_{ij}$, with δ_{ij} being the Kronecker delta, since a walk starting at node i cannot be at the same time at node j unless $i = j$. The network coverage can be approximated by double averaging over all nodes the probability $1 - \sigma_{ij}(t)$:

$$\rho(t) = 1 - \frac{1}{N^2} \sum_{\substack{i,j=1 \\ i\neq j}}^{N} \exp\left[-\mathbf{p}_j(0)\mathbb{P}\mathbf{E}_i^\dagger\right]. \tag{8.23}$$

The agreement between the theoretical values of $\rho(t)$ and Monte Carlo simulation has been shown to be remarkable [54]. Numerical experiments show that, depending on network topologies and transition rules, nodes are visited at different time scales, thus delaying or accelerating the exploration of the multi-layer network when compared to the exploration of each layer separately. This is a genuine multi-layer effect emerging from the interplay between structure and dynamics: coupling different layers might enhance the navigability [54] of the overall system (Figure 8.7 presents an example).

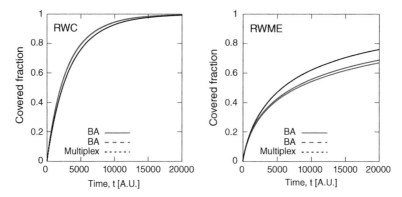

Figure 8.7   Different types of diffusion characterize different topological structures and navigation strategies. Coverage versus time for two different multiplex topologies consisting of two Barabási-Albert (BA) layers (top panels) and a BA layer coupled to a Watts–Strogatz (WS) one (bottom panels). Two different transition rules are considered: RWC (left panels) and RWME (right panels). While the diffusion on single layers separately and on the multiplex are similar for RWC on BA+BA, this is not the case for RWME on BA+BA, where enhanced diffusion is shown in the multiplex. In the other cases, the diffusion is *infra*-diffusive. Figure from De Domenico et al.[54].

8.4   Applications

The interplay between the architecture of a complex network and the dynamics of its units is responsible for the emergence of a variety of interesting phenomena. Here, we briefly review some applications of the network paradigm to case studies of interest in molecular biology.

8.4.1   Identifying Essential Units for the Information Flow

Complex networks have been successfully adopted to model molecular interactions, from genetics [76] to metabolomics [77]. For instance, in the case of protein–protein interactions, it has been shown that a strict relationship between the topological role of a protein and the lethality of its deletion exists [78]. This result, known as the centrality-lethality rule, has been challenged [79].

Only recently have multi-layer networks been considered as models for the proteome of organisms [6, 75, 80]. In this case, layers encode different types of interactions, such as physical and genetic ones (Figure 8.8).

Remarkably, the analysis of a node's PageRank centrality[2] in each layer separately, the corresponding multi-layer network and its aggregate representation might provide, in general, very different results, with the multi-layer centrality providing more reliable results [63]. The annular visualization (Figure 8.9) has been developed to compare the resulting measures, which build multidimensional arrays to be inspected.

---

[2] Note that classical PageRank, as well as its multi-layer generalization [63], can be interpreted as the stationary state of a peculiar random walk dynamics.

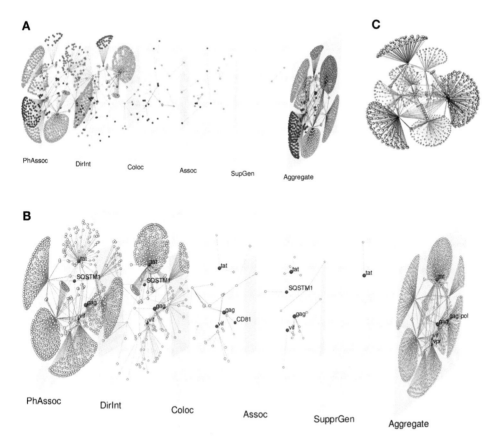

Figure 8.8   Multi-layer network model of protein–protein interactions in HIV-1. Visual representation of the molecular network, with nodes colored by their mesoscale structure (a). Most central nodes are highlighted, and they differ from most central nodes in the aggregate representation of the system (b). Edge-colored visualization of the network (c). Layer–layer correlations (D–E) Figure readapted from [75]. Copyright © 2014, Oxford University Press.

In one case, one can compare different centrality descriptors calculated from the same network model (e.g. the multiplex one), or one can fix a centrality measure and compare the results from different network models.

Understanding how the topological role of molecular units in multi-layer network models is related to their biological function is still an open challenge.

8.4.2   Using Dynamics to Unravel the Mesoscale Structure

Dynamics on the top of the network – more specifically, diffusion dynamics – can be used to identify the organization of nodes in clustered sub-structures, also known as network communities [83, 84]. These clusters aim at modeling the mesoscale structure of the system – by providing a coarse-grain representation – and their detection is important because they can allow one to identify modules with a biological function.

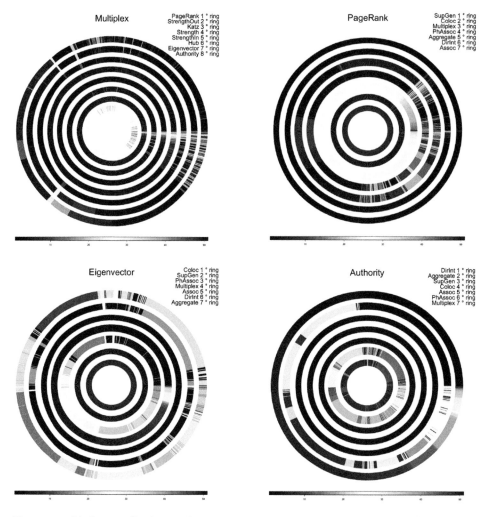

Figure 8.9   Node centrality in protein–protein interactions in HIV-1. Annular visualization of protein centralities calculated for each layer separately, the multi-layer network and its aggregate representation. This visualization allows for a direct comparison of multidimensional results obtained from different approaches, highlighting the differences. Figure readapted from [75]. Copyright © 2014, Oxford University Press.

The underlying idea is that information flowing in the network is expected to be perturbed by clusters [60, 85, 86]. For instance, in the case of random walk dynamics, a random walker will tend to remain trapped for a longer time inside clusters and this effect can be exploited to efficiently identify functional modules. In another approach, the statistical properties of a Markov process taking place on the graph can be used for partitioning the network at multiple time scales.

The same principles have been recently extended to the realm of multi-layer networks in order to identify clusters within and across layers [61, 81]. As an example, we show in Figure 8.10 the analysis of communities in the human proteome according to the Multiplex Infomap method [81]. The multi-layer representation of the human

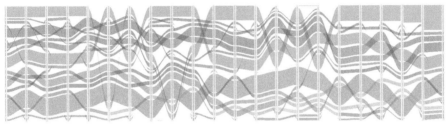

0   0.01 0.06 0.11 0.16 0.21 0.26 0.31 0.36 0.41 0.46 0.51 0.56 0.61 0.66 0.71 0.76 0.81 0.86 0.91 0.96 1.00
Relax rate

Figure 8.10   Changes in the mesoscale organization of the human multiplex proteome for varying relax rate. Clusters, detected by the Multiplex Infomap algorithm [81], with at least 100 proteins, are considered for clarity. This alluvial plot shows how partitions split and merge for increasing rate: larger clusters are quite stable, highlighting that differences in meso-scale are mostly due to smaller sets of proteins. © 2020 IEEE. Reprinted, with permission, from [82].

proteome is an edge-colored network – that is, no direct information about inter-layer connectivity is available. In this case, methods such as the multislice modularity maximization or the Multiplex Infomap have to introduce a parameter to account for the coupling among layers. Such a parameter is called the "relax rate" in the approach based on the Infomap algorithm and it encodes the rate at which random walkers switch layers to explore the multi-layer network. In general, the relax rate is a free parameter. Figure 8.10 shows the changes in the mesoscale organization as a function of the relax rate: some clusters are very stable, but some other ones tend to split and merge for increasing values of the parameter. Very recently, a method to identify a value of the relax rate for which the corresponding mesoscale structure is the most representative, from an information-theoretic perspective, has been proposed [82]. In fact, this goal is achieved by using an independent cost function named the *normalized information loss* (NIL), which can be interpreted as a log-likelihood of a *stochastic block model*. The NIL for a specific relax rate r is defined by

$$H_r(X|Y) = \log_2 \left[\prod_{i=1}^{m} \prod_{j=1}^{m} \binom{n_i n_j}{l_{ij}} \binom{w_{ij}-1}{l_{ij}-1} \right],$$

(8.24)

where n_i indicates the number of nodes in the ith cluster, l_{ij} and w_{ij} the number of links and the total weight of links between clusters i and j. For practical applications, its normalized version

$$H_r^*(X|Y) = \frac{H_r(X|Y) - \min_{0<r\leq1} H_r(X|Y)}{\max_{0<r\leq1} H_r(X|Y) - \min_{0<r\leq1} H_r(X|Y)}$$

(8.25)

is used. By minimizing the NIL, it is possible to learn the latent block structure in the presence of real-valued weights through the use of a parametric distribution, as shown in [87], and the optimal compression of the system is achieved.

It has been shown that clusters identified in the multiplex representation of the human proteome – for the optimal value of the relax given by NIL – have a high functional content when enriched through a standard enrichment-based

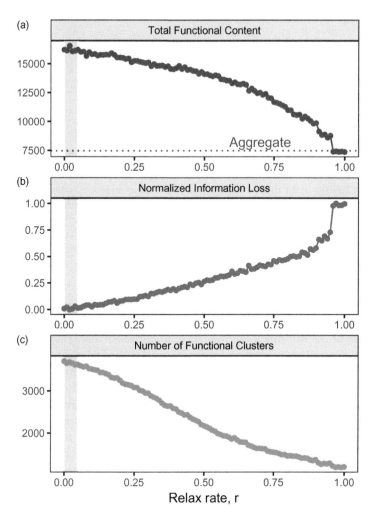

Figure 8.11   Analysis of the mesoscale organization of genes in the human proteome. (a) Total functional content of the cluster collections as a function of the relax rate, computed as the sum of the Jaccard indices of the intersections between each cluster and each gene set in MSigDB (see the text for details). The content is monotonically decreasing for increasing relax rate: for $r \approx 1$, the functional content coincides with the functional content of the cluster collection obtained from the aggregate representation of the system. (b) Normalized information loss (Eq. 8.25) for increasing relax rate, highlighting an increasing degradation of the information content. (c) Decreasing number of identified functional clusters for increasing relax rate. In all panels, the shaded area highlights the range of relax rates where normalized information loss is minimum which, remarkably, coincides with the range where the biological enrichment of identified clusters exhibits maximum functional content. © 2020 IEEE. Reprinted, with permission, from [82].

strategy using the Molecular Signatures Database[3] (MSigDB), as shown in Fig. 8.11, making this method a desirable tool for the analysis of complex multi-layer biological systems.

---

[3] http://software.broadinstitute.org/gsea/msigdb/collections.jsp.

8.5   Conclusions

A variety of complex systems, including biological ones, consist of entities interacting with each other in multiple ways. The topology of most biological molecular networks and the dynamics of their units are responsible for cell function. Multi-layer network models and analysis provide a promising framework to cope with this new level of complexity and to improve our understanding of such complex systems.

Cells' molecular essential components are functionally interdependent and related by interactions of different types (e.g. genetic, physical, etc.) at different scales (e.g. genetic, metabolic, etc.). Therefore, multi-layer network modeling is expected to play an essential role for their modeling and analysis.

The study of biological systems from a multi-layer perspective is relevant because of its deep implications for life and disease. In fact, even small perturbations to single units (e.g., a single gene or one specific metabolite) can quickly propagate within each system and across scales, causing abnormal functions in tissues and organs that culminate in diseases. While network medicine [88] has proven to be an effective tool for better analyzing and understanding the relationship between molecular biological systems and phenotypes, multi-layer network approaches similar in spirit [89] are still not fully developed, providing exciting research opportunities for the future.

References

[1] Buldyrev SV, Parshani R, Paul G, Stanley HE, Havlin S. Catastrophic cascade of failures in interdependent networks. *Nature*. 2010;464(7291):1025–1028.

[2] Gao J, Buldyrev SV, Havlin S, Stanley HE. Robustness of a network of networks. *Phys Rev Lett*. 2011;107(19):195701.

[3] Cellai D, López E, Zhou J, Gleeson JP, Bianconi G. Percolation in multiplex networks with overlap. *Phys Rev E*. 2013;88(5):052811.

[4] Bianconi G and Dorogovtsev SN. Multiple percolation transitions in a configuration model of a network of networks. *Phys Rev E*. 2014;89(6):062814.

[5] Min B, Do Yi S, Lee K-M, Goh K-I. Network robustness of multiplex networks with interlayer degree correlations. *Phys Rev E*. 2014;89(4):042811.

[6] Radicchi F. Percolation in real interdependent networks. *Nat Phys*. 2015;11(7):597.

[7] Hackett A, Cellai D, Gómez S, Arenas A, Gleeson JP. Bond percolation on multiplex networks. *Phys Rev X*. 2016;6(2):021002.

[8] Radicchi F and Bianconi G. Redundant interdependencies boost the robustness of multiplex networks. *Phys Rev X*. 2017;7(1):011013.

[9] Osat S, Faqeeh A, Radicchi F. Optimal percolation on multiplex networks. *Nat Commun*. 2017;8(1):1540.

[10] Yagan O and Gligor V. Analysis of complex contagions in random multiplex networks. *Phys Rev E*. 2012;86(3):036103.

[11] Brummitt CD, Lee K-M, Goh K-I. Multiplexity-facilitated cascades in networks. *Phys Rev E*. 2012;85(4):045102.

[12] Cozzo E, Banos RA, Meloni S, Moreno Y. Contact-based social contagion in multiplex networks. *Phys Rev E*. 2013;88(5):050801.

[13] Hu Y, Havlin S, Makse HA. Conditions for viral influence spreading through multiplex correlated social networks. *Phys Rev X*. 2014;4(2):021031.

[14] Lima A, De Domenico M, Pejovic V, Musolesi M. Disease containment strategies based on mobility and information dissemination. *Sci Rep*. 2015;5:10650.

[15] Stella M, Andreazzi CS, Selakovic S, Goudarzi A, Antonioni A. Parasite spreading in spatial ecological multiplex networks. *J Complex Netw*. 2016;5(3):486–511.

[16] Gómez-Gardenes J, Reinares I, Arenas A, Floría LM. Evolution of cooperation in multiplex networks. *Sci Rep*. 2012;2.

[17] Wang Z, Szolnoki A, Perc M. Interdependent network reciprocity in evolutionary games. *Sci Rep*. 2013;3:1183.

[18] Matamalas JT, Poncela-Casasnovas J, Gómez S, Arenas A. Strategical incoherence regulates cooperation in social dilemmas on multiplex networks. *Sci Rep*. 2015;5:srep09519.

[19] Battiston F, Perc M, Latora V. Determinants of public cooperation in multiplex networks. *New J Phys*. 2017;19:071002.

[20] Battiston F, Nicosia V, Latora V, Miguel MS. Layered social influence promotes multiculturality in the Axelrod model. *arXiv:1606.05641*, 2016.

[21] Battiston F, Cairoli A, Nicosia V, Baule A, Latora V. Interplay between consensus and coherence in a model of interacting opinions. *Physica D*. 2016;323:12–19.

[22] Solé-Ribalta A, De Domenico M, Kouvaris NE, et al. Spectral properties of the Laplacian of multiplex networks. *Phys Rev E*. 2013;88(3):032807.

[23] Aguirre J, Sevilla-Escoboza R, Gutiérrez R, Papo D, Buld J. Synchronization of interconnected networks: the role of connector nodes. *Phys Rev Lett*. 2014;112(24):248701.

[24] Gambuzza LV, Frasca M, Gomez-Gardeñes J. Intra-layer synchronization in multiplex networks. *EPL*. 2015;110(2):20010.

[25] Zhang X, Boccaletti S, Guan S, Liu Z. Explosive synchronization in adaptive and multilayer networks. *Phys Rev Lett*. 2015;114(3):038701.

[26] Del CI Genio, Gómez-Gardeñes J, Bonamassa I, Boccaletti S. Synchronization in networks with multiple interaction layers. *Sci Adv*. 2016;2(11):e1601679.

[27] Asllani M, Busiello DM, Carletti T, Fanelli D, Planchon G. Turing patterns in multiplex networks. *Phys Rev E*. 2014;90(4):042814.

[28] Kouvaris NE, Hata S, A Díaz-Guilera. Pattern formation in multiplex networks. *Sci Rep*. 2015;5:srep10840.

[29] Nicosia V, Bianconi G, Latora V, Barthelemy M. Growing multiplex networks. *Phys Rev Lett*. 2013;111:058701.

[30] Kim JY and Goh K-I. Coevolution and correlated multiplexity in multiplex networks. *Phys Rev Lett*. 2013;111(5):058702.

[31] Gao J, Buldyrev SV, Stanley HE, Havlin S. Networks formed from interdependent networks. *Nat Phys*. 2012;8(1):40–48.

[32] Kivelä M, Arenas A, Barthelemy M, et al. Multilayer networks. *J Complex Netw*. 2014;2(3):203–271.

[33] Boccaletti S, Bianconi G, Criado R, et al. The structure and dynamics of multilayer networks. *Phys Rep*. 2014;544(1):1–122.

[34] Wang Z, Wang L, Szolnoki A, Perc M. Evolutionary games on multilayer networks: a colloquium. *Eur Phys J B*. 2015;88:124.

[35] De Domenico M, Granell C, Porter MA, Arenas A. The physics of spreading processes in multilayer networks. *Nat Phys*. 2016;12(10):901–906.

[36] Battiston F, Nicosia V, Latora V. The new challenges of multiplex networks: measures and models. *Eur Phys J Special Top*. 2017;226(3):401–416.

[37] Dickison M, Havlin S, Stanley HE. Epidemics on interconnected networks. *Phys Rev E*. 2012;85(6):066109.

[38] Sanz J, Xia C-Y, Meloni S, Moreno Y. Dynamics of interacting diseases. *Phys Rev X*, 4(4):041005.

[39] Ferraz de Arruda G, Cozzo E, Rodrigues FA, Moreno Y. Epidemic spreading in interconnected networks: a continuous time approach. *arXiv:1509.07054*, 2015.

[40] Wang Z, Andrews A, Wang L, Bauch CT. Coupled disease–behavior dynamics on complex networks: a review. *Phys Life Rev*. 2015;15:1–29.

[41] Funk S, Bansal S, Bauch CT, et al. Nine challenges in incorporating the dynamics of behaviour in infectious diseases models. *Epidemics*. 2015;10:21–25.

[42] Funk S, Gilad E, Watkins C, Jansen VAA. The spread of awareness and its impact on epidemic outbreaks. *PNAS*. 2009;106(16):6872–6877.

[43] Granell C, Gómez S, Arenas A. Dynamical interplay between awareness and epidemic spreading in multiplex networks. *Phys Rev Lett*. 2013;111:128701.

[44] Granell C, Gómez S, Arenas A. Competing spreading processes on multiplex networks: awareness and epidemics. *Phys Rev E*. 2014;90(1):012808.

[45] Nicosia V, Skardal PS, Arenas A, Latora V. Collective phenomena emerging from the interactions between dynamical processes in multiplex networks. *Phys Rev Lett*. 2017;118(13):138302.

[46] De Domenico M. Multilayer network modeling of integrated biological systems. *Phys Life Rev*. 2018;24:149–152.

[47] Radicchi F, Arenas A. Abrupt transition in the structural formation of interconnected networks. *Nat Phys*. 2013;9:717–720.

[48] Radicchi F. Driving interconnected networks to supercriticality. *Phys Rev X*. 2014;4(2):021014.

[49] Chung FRK. *Spectral Graph Theory*. 2nd edition. Providence, RI: American Mathematical Society, 1997.

[50] De Domenico M, Solé-Ribalta A, Cozzo E, et al. Mathematical formulation of multilayer networks. *Phys Rev X*. 2013;3(4):041022.

[51] Gómez S, Diaz-Guilera A, Gómez-Gardenes J, et al. Diffusion dynamics on multiplex networks. *Phys Rev Lett*. 2013;110(2):028701.

[52] Kolda TG, Bader BW. Tensor decompositions and applications. *SIAM Rev*. 2009;51(3):455–500.

[53] Bazzi M, Porter MA, Williams S, et al. Community detection in temporal multilayer networks, with an application to correlation networks. *Multiscale Model Simul*. 2016;14(1):1–41.

[54] De Domenico M, Solé-Ribalta A, Gómez S, Arenas A. Navigability of interconnected networks under random failures. *PNAS*. 2014;111(23):8351–8356.

[55] Chung FRK. *Spectral Graph Theory*. 2nd edition.Providence, RI: American Mathematical Society, 1997.

[56] Noh JD and Rieger H. Random walks on complex networks. *Phys Rev Lett*. 2004;92(11):118701.

[57] Newman M. *Networks: An Introduction*. New York: Oxford University Press, 2010.

[58] Masuda N, Porter MA, Lambiotte R. Random walks and diffusion on networks. *Phys Rep*. 2017;716–717:1–58.

[59] Brin S, Page L. The anatomy of a large-scale hypertextual web search engine. *Comput Netw ISDN Syst*. 1998;30(1):107–117.

[60] Rosvall M, Bergstrom CT. An information-theoretic framework for resolving community structure in complex networks. *PNAS*. 2007;104(18):7327–7331.

[61] Mucha PJ, Richardson T, Macon K, Porter MA, Onnela J-P. Community structure in time-dependent, multiscale, and multiplex networks. *Science*. 2010;328(5980):876–878.

[62] S-Yang J. Exploring complex networks by walking on them. *Phys Rev E*. 2005;71(1):016107.

[63] De Domenico M, Solé-Ribalta A, Omodei E, Gómez S, Arenas A. Ranking in interconnected multilayer networks reveals versatile nodes. *Nat Commun*. 2015;6.

[64] Solé-Ribalta A, De Domenico M, Gómez S, Arenas A. Centrality rankings in multiplex networks. In *Proceedings of the 2014 ACM Conference on Web Science*. New York: ACM, 2014, pp. 149–155.

[65] Solé-Ribalta A, De Domenico M, Gómez S, Arenas A. Random walk centrality in interconnected multilayer networks. *Physica D*. 2016;323:73–79.

[66] Lovász L. Random walks on graphs: A survey. *Combinator, Paul Erdos Eighty*. 1993;2(1):1–46.

[67] Zhang Z, Julaiti A, Hou B, Zhang H, Chen G. Mean first-passage time for random walks on undirected networks. *Euro Phys J B*. 2011;84(4):691–697.

[68] Kemeny JG, Snell JL. *Finite Markov Chains*. New York: Van Nostrand Reinhold, 1960.

[69] Wilson RJ. *Introduction to Graph Theory*. New York: Academic Press, 1972.

[70] Tetali P. Random walks and the effective resistance of networks. *J Theoret Prob*. 1991;4(1):101–109.

[71] Samukhin A, Dorogovtsev S, Mendes J. Laplacian spectra of, and random walks on, complex networks: are scale-free architectures really important? *Phys Rev E*. 2008;77(3):036115.

[72] Burda Z, Duda J, Luck J, Waclaw B. Localization of the maximal entropy random walk. *Phys Rev Lett*. 2009;102(16):160602.

[73] Watts D, Strogatz S. Collective dynamics of "small-world" networks. *Nature*. 1998;393(6684):440–442.

[74] Battiston F, Nicosia V, Latora V. Efficient exploration of multiplex networks. *New J Phys*. 2016;18(4):043035.

[75] De Domenico M, Porter MA, Arenas A. Muxviz: a tool for multilayer analysis and visualization of networks. *J Complex Netw*. 2015;3(2):159.

[76] Costanzo M, Baryshnikova A, Bellay J, et al. The genetic landscape of a cell. *Science*. 2010;327(5964):425–431.

[77] Guimera R, Amaral LAN. Functional cartography of complex metabolic networks. *Nature*. 2005;433(7028):895.

[78] Jeong H, Mason SP, Barabási A-L, ZN Oltvai. Lethality and centrality in protein networks. *Nature*. 2001;411(6833):41.

[79] He X, Zhang J. Why do hubs tend to be essential in protein networks? *PLoS Genet*. 2006;2(6):e88.

[80] De Domenico M, Nicosia V, Arenas A, Latora V. Structural reducibility of multilayer networks. *Nat Commun*. 2015;6:6864.

[81] De Domenico M, Lancichinetti A, Arenas A, Rosvall M. Identifying modular flows on multilayer networks reveals highly overlapping organization in interconnected systems. *Phys Rev X*. 2015;5(1):011027.

[82] Mangioni G, Jurman G, De Domenico M. Multilayer flows in molecular networks identify biological modules in the human proteome. *IEEE Trans Network Sci Eng*. 2020;7(1):411–420.

[83] Newman ME. Communities, modules and large-scale structure in networks. *Nat Phys*. 2012;8(1):25.

[84] Fortunato S, Hric D. Community detection in networks: a user guide. *Phys Rep*. 2016;659:1–44.

[85] Gfeller D, De Los Rios P. Spectral coarse graining of complex networks. *Phys Rev Lett*. 2007;99(3):038701.

[86] Lambiotte R, Delvenne J-C, Barahona M. Random walks, Markov processes and the multiscale modular organization of complex networks. *IEEE Trans Network Sci Eng*. 2014;1(2):76–90.

[87] Aicher C, Jacobs AZ, Clauset A. Learning latent block structure in weighted networks. *J Complex Netw*. 2015;3(2):221–248.

[88] Barabási A-L, Gulbahce N, Loscalzo J. Network medicine: a network-based approach to human disease. *Nat Rev Genet*. 2011;12(1):56.

[89] Klosik DF, Grimbs A, Bornholdt S, Hütt M-T. The interdependent network of gene regulation and metabolism is robust where it needs to be. *Nat Commun*. 2017;8(1):534.

PART IV
APPLICATIONS

9 The Network of Networks Involved in Human Disease

Celine Sin and Jörg Menche

9.1 Introduction

Human health and disease are influenced by an astounding complexity of intertwined processes that span many orders of magnitude in both space and time (Figure 9.1). At the molecular level, for example, the five primary nucleobases that form the basic units of the genetic code are relatively simple molecules consisting of only 12–16 atoms. Yet, lined up end to end, the human genome stretches over 3 meters in length, and it is these same nucleotides that encode all the processes of life from the simplest single-celled organisms up to the most complex animals. Some of the most fundamental chemical processes of life, such as the conversion of biochemical energy from nutrients into adenosine triphosphate (ATP), occur at the order of microseconds. On the other hand, decade-long processes such as malnutrition or environmental exposures are key to understanding some of the most common diseases, such as heart diseases or diabetes.

Over the last years, it has been increasingly recognized that the reductionist approach – investigating proteins or individual organs in isolation – is rarely sufficient for a complete understanding of pathobiological or potential therapeutic approaches [1]. Indeed, the many components of biological systems, from molecules to cells to complex organisms, interact in an intricate and tightly coordinated fashion. Disease states can be understood as perturbations out of balance from normal function, either from outside sources or abnormalities within the components or their interactions. While some diseases can be traced back to a single genetic defect, or invasion by specific microorganisms, such as viruses or bacteria, others develop from the combined effect of multiple factors. Contrary to the archetype of "one-gene, one-function, one-phenotype" proposed in the 1940s [2], it is increasingly evident that a particular gene abnormality not only impacts the activity of its particular gene product, but also trickles through the whole cascade of interactions between several sub- to supra-cellular processes [3]. In other words, a disease phenotype is rarely a consequence of an abnormality in a single gene, but reflective of a slew of perturbed pathobiological processes that interact in complex networks.

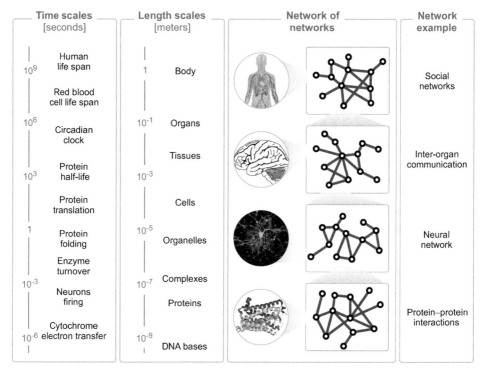

Figure 9.1   The network of networks in medicine. The processes involved in health and disease span a wide range of scales in both time and space. Different processes are relevant at different scales. These processes can be summarized by networks in which the actors are represented by nodes and their relationships as links. An entire network at one scale may be embedded as a single node into the next.

As the components relevant to health and disease span many orders of magnitudes, so do their interactions. We can conceptualize these multi-scale interactions as an embedding of networks into networks [4] (Figure 9.1). Proteins interact with each other in our cells, which in turn interact with other cells forming whole organs, and so on, all the way to social interactions among humans. Over the last two decades, these networks have been identified and mapped more and more accurately. Network science plays a crucial role in solving the next part of the puzzle, which is how to interpret and ultimately understand the functional, logical and dynamic aspects of how they contribute to diverse disease phenomena [5, 6]. This endeavor is inherently cross- and interdisciplinary, and network science can rely on a rich body of rigorous results from graph theory and statistical physics, as well as tools and concepts from computer science and sociology.

   In the following, we will first give an overview of the most important biological network layers and their uses in biomedical research. We will then provide more details on how the emerging field of network medicine uses interactome networks as maps to study human disease. Finally, we will introduce a network-based methodology for elucidating the molecular mechanisms of specific diseases.

9.2   The Networks of Networks in Health and Disease

We can broadly classify the multitude of networks relevant in biomedical research into two main categories: (1) networks based on direct physical interactions and thereby representing the innate organization of many biological systems, such as the network of physical protein interactions or cell–cell contacts among neurons; (2) networks that serve as an analysis framework to investigate more abstract relationships, such as shared molecular mechanisms among diseases. With the increasing sophistication of -omic data collection, growing knowledge of biological processes, and access to population-wide health records, we are in an era of explosive data growth. In addition to being "big," these data are often noisy, and interpretation is often not straightforward. Networks provide a convenient and powerful toolbox to systematically analyze these complex systems in a comprehensive and holistic fashion.

9.2.1   Networks as Organizational Principle of Biological Systems

Following roughly the length scales of biological organization outlined in Figure 9.1 from smallest to largest, we start by discussing important molecular networks.

Metabolic Networks
Among the most comprehensive and best-studied physical interaction networks are metabolic networks. Metabolism is the collection of all the chemical processes in a cell, such as the conversion of one metabolite to the next. These processes provide both the energy and the building blocks that are essential for all life. Metabolic networks represent an integrated map of all these processes with metabolites as nodes (e.g. glucose, ATP, triglycerol), which are linked through reactions or enzymes converting one metabolite to the other. There are several curated databases of metabolic maps, such as the Human Recon 2.2 [7], containing 5324 metabolites, 7785 reactions and 1675 associated genes; the Edinburgh human metabolic network [8], with 3000 reactions and over 2000 associated genes; and the Kyoto Encyclopedia of Genes and Genomes (KEGG) [9], which also contains data on a wide range of other species.

Constructing context-dependent metabolic networks for different cell types or under different perturbations is particularly useful for the identification of essential pathways and prediction of cellular responses to different treatments. For example, a metabolic network of the human hepatocyte showed how the liver responds to the availability of different metabolites in order to maintain homeostasis in the blood [10].

Network analysis of metabolic networks from disease states is also informative to elucidate pathologies that result from deficiencies or excessive amplification of metabolic pathways. In [11], metabolic networks were supplemented with gene expression data to identify principle metabolic and regulatory nodes in type 2 diabetes mellitus. For different cancer cell lines, metabolic networks have been used to predict drug efficacy [12], and patient survivability [13].

Gene Regulatory Networks
Another essential dynamical process requiring tight regulation is gene expression. Compared to metabolism, our knowledge of gene regulatory processes is much

less complete. The most basic nodes in gene regulatory networks are transcription factors and their respective target DNA regulatory elements [14]. Network motifs can reveal important principles of regulatory mechanisms (e.g. autoregulation, feedforward loops) [15]. Commonly used databases containing experimentally verified genetic regulatory interactions include JASPAR [16] and TRANSFAC [17]. However, gene expression is not only regulated through transcription factors, but also through other interactions, such as between RNAs or RNA and DNA. For example, microRNA plays a considerable role in regulating mRNA concentrations by modulating mRNA stability and degradation [18]. Additionally, microRNA was shown to regulate pre-mRNA processing in the nucleus, assist in mRNA structure formation, and modulate mRNA–protein interactions [19]. Numerous microRNA are associated with the development of several diseases [20, 21]. Interactions between RNA and/or DNA species can be measured experimentally [22, 23] or predicted computationally [24, 25]. Databases containing computationally predicted interactions include TargetScan [25], PicTar [26], microRNA [27], miRWalk [28] and miRBase [29]; experimentally confirmed interactions are collected in TarBase [30] miRecords [31].

Protein–Protein Interaction Networks

The gene products that emerge from the regulatory process then engage in numerous physical interactions with each other to perform a multitude of molecular processes in the cell. Such protein–protein interactions are in general very specific, mediated through complementary "lock and key" interaction interfaces. As these interactions are integral to normal cell functioning, mutations affecting the interaction interfaces often have particularly strong effects.

Two main experimental techniques allow for systematic, large-scale mapping of protein–protein interactions: binding affinity purifications coupled to mass spectrometry [32, 33] and yeast two-hybrid (Y2H) assays [6, 34]. In the first method, cell lines are engineered to produce a "bait" protein with a tag that can be captured by beads. These proteins (and their interaction partners) are then captured and identified through mass spectrometry. Since often a whole complex is collected, translation of the data into direct pairwise interactions is often difficult. Yet, the interactions revealed by this method are specific to a particular biologically relevant condition. On the other hand, Y2H assays map out precise binary protein interactions. However, not all interactions found are biologically relevant; for example, two respective interacting proteins might never be expressed at the same time in the same cell [35]. Additionally, there are many small-scale studies using a myriad of methods, such as co-immunoprecipitation, X-ray crystallography or nuclear magnetic resonance. Computational predictions of protein interactions use features of amino acid sequences [36–39], gene fusion [40], or phylogenetic trees [41]. Each of these sources of protein interactions have strengths and limitations in terms of comprehensiveness, noise and biases [42], such as biases in the selection of protein pairs [43] or experimental biases, for example toward highly expressed genes [34].

There are several online depositories of protein–protein interaction data. EMBL-EBI maintains the manually curated "IntAct" molecular interaction database containing 57 857 proteins and 275 145 interactions from over 5000 publications [44, 45].

"Reactome" is another manually curated database containing also interactions between protein and nucleic acids, small molecules and macromolecules [46, 47]. The Human Integrated Protein Protein Interaction rEference (HIPPIE) collects data from several primary databases and offers a confidentiality score for each reported interaction [48]. The Search Tool for Recurring Instances of Neighboring Genes (STRING) database contains even more interactions by including predictions, for example based on co-expression or automated text mining [49].

Protein interaction networks have found numerous applications, ranging from elucidating basic principles of cellular organization [50] to the prediction of disease genes [51–53] or of the therapeutic effect of drugs [54] (see Sections 9.3 and 9.4).

Neuronal Networks

The networks considered so far occur, for the most part, within individual cells. An important example of inter-cellular networks in which different cells communicate with each other are networks of neurons. Neurons exchange signals either chemically, through the release of neurotransmitters at synapses, or electrically via the flow of ions through gap junctions. The first, and so far only, completely mapped network of all neural connections of an organism was published as early as 1986 for the roundworm *C. elegans* and contains around 9000 chemical and around 800 electrical connections between around 300 neurons [55]. Partial neural networks are available, for example, for mice [56, 57], and there are also first, very coarse-grained data sets available for humans [58]. The ultimate goal of mapping the complete human "connectome" of all our brain cells will likely remain out of reach for many years to come [59].

Network analysis of these connectomes has contributed to our understanding of basic principles of neural function [60] and cognition [61], as well as psychopathologies [62].

The Immune System

Similar to the nervous system that interfaces to many other body parts, but perhaps even more diverse in terms of participating organs, cell types and molecules, is the immune system. To meet the constant challenges of internal and external threats, ranging from tumor cells to bacterial infections, our immune system orchestrates a multitude of cells with often highly specialized functions. The nodes in the "social network" of the immune system therefore represent cells, links represent communication through signaling molecules, such as cell-surface receptors or secreted molecules [63]. This network can be mapped using mass spectrometry by measuring the proteomes of immune cell populations and comparing the levels of intracellular and secreted proteins between different stimuli. A recent study identified over 180 000 high-confidence interactions between 460 receptors and 300 ligands in such a fashion [64]. A network analysis revealed several principles of intra-cellular communication in the immune system. For example, lineages that are developmentally less related to each other tend to have a higher number of interactions and different immune cells exhibit pronounced differences in their communication patterns after being activated. Immune networks constructed from large-scale text mining have also been used to predict cytokine disease associations [65].

Population Networks

While the "social network" served merely as a metaphor in the case of the immune system, the actual relationships between humans are also important for a number of diseases, most notably for the spread of contagious diseases, such as viral or bacterial infections. The first mathematical models of diseases spreading among individuals of a population were formulated in 1760 by the Swiss mathematician and physicist Daniel Bernoulli [66]. As maps of the networks on which diseases propagate become available, in particular global transportation maps and networks of social interactions, these models are becoming increasingly accurate [67]. Network-based epidemiological models can help us understand global propagation patterns observed in recent pandemic outbreaks, identify the source of an outbreak, predict future highly affected areas or design effective immunization or prevention strategies [68, 69].

Interestingly, the spread of diseases through social contacts is not only limited to diseases that are transmitted through viruses or bacteria. It has been shown that also obesity [70], the tendency to start to smoke [71] or general happiness in life [72] may propagate along social connections between people.

9.2.2   Networks as Data Analysis Tools

All networks reviewed above build on direct, often physical relationships between entities ranging from molecules to people. However, networks can also be used to characterize more abstract, less direct relationships. In the following we introduce important examples of such networks that represent non-physical, yet still biologically highly relevant relationships.

Co-expression Networks

As mentioned above, our knowledge of gene regulatory networks remains scarce as they are highly context-dependent, involve a large number of diverse molecules and are therefore difficult to assess experimentally in a comprehensive fashion. A much more easily accessible quantity that may serve as a proxy to study gene regulatory programs is co-expression. Two genes are co-expressed if their respective expression levels correlate strongly under different experimental conditions, such as over time or under different stimuli. Genome-wide gene expression can be assessed using RNAseq technology, enabling the construction of large-scale co-expression networks [73, 74]. A comprehensive database of expression data across many tissues, cell types and conditions is collected and curated by the GTEx consortium [75]. In contrast to gene regulatory networks, co-expression networks do not imply a causal relationship between genes. Yet, co-expression networks can still be used to identify groups of genes that are more broadly functionally related – for example controlled by the same transcriptional regulatory program, or members of the same pathway or protein complex [76]. Analyses of these networks have identified commonly affected pathways in autism spectrum disorder [77], Alzheimer's disease [78] and inflammatory bowel disease [79]. Candidate biomarkers for myocardial infarction [80] and several cancers [81, 82] have also been identified using co-expression networks.

Genetic Interaction Networks

Another important indirect relationship between genes is given by genetic interactions. Most generally, these interactions describe the phenomena of observing an unexpected phenotype upon simultaneous mutations in two genes. More specifically, two genes are said to have a negative genetic interaction if mutations in the genes individually are not lethal, but become lethal when simultaneously mutated. Conversely, genes are said to have a positive genetic interaction if a mutation in one gene "rescues" a lethal mutation in another [83].

Genetic interactions can be evaluated by creating gene-deletion mutants for the genes of interest. Large-scale screens have been performed in yeast [83–85]. The most comprehensive screen in human haploid cells identified approximately 2000 essential genes, revealed genes regulating the secretory pathway and generated new insights into Golgi apparatus homeostasis [86]. There are also several more specialized screens in human cells, focusing on tumor suppressor genes [87, 88] and cancer drug targets [89], for example.

In addition to studying the large-scale functional organization of genes, genetic interactions also hold great promise for concrete therapeutic applications. For example, a positive genetic interaction with the BLM helicase complex was recently shown to rescue the Fanconi anemia (FA) phenotype caused by a loss of function mutation in the FA gene that leads to defective DNA damage repair [90]. Furthermore, genetic interaction networks also hold special promise for studying complex diseases, such as cancer, that result from a number of genetic mutations (and environmental factors, potentially) that impact several subcellular systems. In [88], the genetic interaction network was used to identify potential chemotherapeutic drug targets.

Co-perturbation Networks

The concept of genetic interactions can be generalized from gene inactivation to arbitrary perturbations. Co-perturbation networks thus encapsulate the information from perturbation biology screens, with nodes representing genes and edges again representing significant correlations among the response of the system toward perturbations in the two respective genes. Examples of such perturbations that are assessed in high-throughput measurements range from RNAi [91] or CRISPR [92] to drug treatment [93–95]. Commonly used readouts of the cellular response include gene expression through RNAseq technology [96] and high-resolution fluorescent microscopy [97]. Co-perturbation networks have been used, for example, to predict drug targets [96], elucidate molecular mechanisms of drugs [98, 99] or infer pathway activity from gene expression [100–102].

Disease Networks

Diseases, while having diverse causes, development and manifestations, often share a number of similar characteristics. These relationships among diseases may occur at several scales and can be systematically investigated using disease networks: on the molecular level (e.g. sharing common genetic origin), on the phenotypic level (e.g. sharing common clinical signs and symptoms) and on the population level (e.g. having frequent co-occurrence in patients). The first comprehensive map of the human "diseaseome" was presented in [103], linking 1377 diseases based on their shared genetic associations as reported in the OMIM database [104]. The resultant network clearly showed that very few diseases could be regarded as isolated entities and directly attributed to a single distinct origin. Instead, the majority of diseases fall into highly connected clusters of disease groups, with overlapping molecular roots. The network properties within and around disease clusters are also predictive of disease characteristics: diseases occupying a more central position in the disease network tend to be more prevalent and have higher mortality rates [105].

Genetic overlap among diseases extends to the physical interactions among the respective gene products, to the resultant gene expression profiles, to the cells, to the organ systems, and eventually to the organism. Thus, it is a logical progression to expect similar findings at the phenotypic scale. In [106], a disease network based on the similarity of clinical symptoms was built using the annotated Medical Subject Headings (MeSH) metadata [107]. Indeed, two diseases sharing similar symptoms tended to also share similar defects on protein interactions, if not the genetic associations directly. The study further revealed that the degree of localization of the associated genes on the underlying protein interaction network is indicative of the diversity of clinical manifestations. That is, diseases that are more localized on the protein interaction network tend to have narrower clinical presentation. Comparison of protein interaction networks and disease networks from different classes of disease (e.g. complex diseases, Mendelian diseases, cancer) revealed interesting differences between diseases with difference inheritance modes [108–110].

Disease relationships at the population level can be evaluated through co-morbidity networks. Co-morbidity describes the tendency of certain diseases to co-occur in the same patient. Such data can be mined from patient records – in [111], a disease network built from 30 million patient records showed a relationship between disease progression patterns to topological properties of the respective disease. Highly central, highly connected diseases are associated with a higher mortality rate, and patients tend to be affected by peripheral diseases before developing more central ones. More recently, similar differences in disease progression patterns related to age and sex have been characterized [112]. Co-morbidity networks have been used to explore possibilities for drug repurposing [113], evaluate potential drug side-effects [114], identify biomarkers [115], and disentangle genetic and environmental factors of diseases [116].

9.3   Molecular Networks as Maps

Despite the large diversity of networks introduced in the previous section, they exhibit certain universal features and patterns (Figure 9.2). (1) The position of an individual node within the network is often related to its importance within the represented biological system. (2) The local connectivity among a group of nodes can be associated with shared (patho-)biological functions. (3) The network distance between groups of nodes often indicates their degree of relatedness. Taken together, these features form the basis of viewing molecular networks as maps. In this section, we review this important metaphor in more detail using the interactome as an example – that is, the integrated network of molecular interactions in the cell. The tools and concepts that we will introduce can be readily applied to other networks as well.

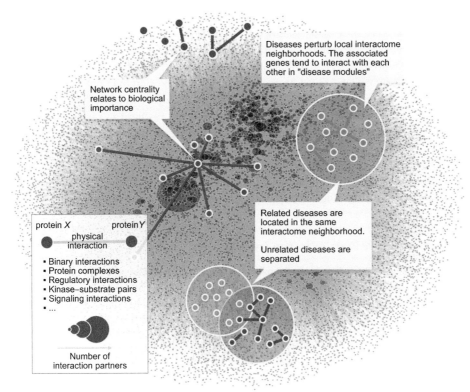

Diseases perturb local interactome neighborhoods. The associated genes tend to interact with each other in "disease modules"

Network centrality relates to biological importance

protein X   protein Y
physical interaction
- Binary interactions
- Protein complexes
- Regulatory interactions
- Kinase–substrate pairs
- Signaling interactions
- ...

Related diseases are located in the same interactome neighborhood.

Unrelated diseases are separated

Number of interaction partners

Figure 9.2   The interactome as a map. The interactome represents all biologically relevant molecular interactions in the cell. The protein–protein interaction network shown here contains 13 460 proteins connected by 141 296 interactions [117]. The annotations on top illustrate the basic findings that inspired the analogy between networks and maps.

9.3.1   Basic Interactome Properties

As introduced above, the various molecular interactomes can be broadly categorized into direct physical and more indirect functional relationships. Focusing on the physical interactions, in particular on protein–protein interactions, we can identify three main sources: (1) interactions curated from the scientific literature, mainly derived from specific, small-scale experiments; (2) interactions from systematic, proteome-scale mapping efforts; and (3) interactions from computational predictions. Figure 9.2 gives a visual impression of a manually curated interactome network focusing only on physical interactions with direct experimental evidence [117]. It contains 13 460 proteins connected by 141 296 interactions between them. On average, each protein has approximately 21 interaction partners, but in this network, as well as in many complex networks, the number of interaction partners per node ("degree," k) vary widely: while the majority of nodes have only a few neighbors (more than 2000 proteins have only one interaction partner), there are also a handful of nodes with hundreds of connections, such as $GRB2$ (degree $k = 872$), $YWHAZ$ ($k = 502$) and $TP53$ ($k = 450$), the so-called "hubs." The heterogeneous distribution of degrees, and in particular the existence of hubs, have a profound effect on many network properties. Hubs connect many distinct parts of the network, shortening the distance between nodes, also known as the "small world effect" [118]. In some cases of scale-free networks whose degree distribution approximates $P(k)\,k^{-\gamma}$, one can even observe "ultra small world effects" [119]. In the interactome, it takes an average of fewer than four hops ($\langle d \rangle = 3.6$) to move from any protein to any other protein. Networks that have such high degree of connectedness tend to be very resilient to random failure – that is, the structure of the network is preserved despite random removal of nodes or edges [120–123]. While these networks are robust against random failure, they are also particularly vulnerable to targeted attacks of the hubs [124]. In the interactome, the removal of the $\sim$30% of the most highly connected nodes is sufficient to completely destroy the network, leaving only disconnected fragments.

9.3.2   Node Localization in the Interactome

Given the vulnerability of such networks to attacks targeted toward highly connected nodes, we expect that proteins that serve as hubs in a biological network also have higher biological importance. Indeed, as first shown in yeast (*Saccharomyces cerevisiae*) [50] and later confirmed in human cell lines [125], the protein products of essential genes – the genes that are crucial for survival – tend to be hubs located toward the center of the interactome. Conversely, less essential genes tend to have fewer interactions and are situated more in the periphery of the interactome. These findings were later extended and refined for disease-associated genes, revealing specific topological properties that differ between classes of diseases (e.g. complex diseases, Mendelian diseases, cancer) and inheritance modes (autosomal dominant or recessive). Cancer-driver genes, for example, are often highly central, while recessive disease genes tend to be located toward the periphery [110].

9.3.3   Neighborhoods in the Interactome

Beyond measures of centrality and connectedness of individual nodes, many important structural connection patterns between a group of nodes have been identified. For example, "network modules," that is nodes that are densely connected among themselves but only sparsely connected to the rest of the network, often perform a certain function together [126–128]. Similarly, shared pathway membership, co-localization in the cell and co-expression [33, 34] have been found to be associated with specific interactome neighborhoods. In addition to functional relationships, network modules have also been identified with disease-related processes, showing that genes implicated in the same disease tend to be more connected to each other than expected by chance [129]. A systematic study on $\sim$300 complex diseases revealed that currently available interactome networks offer sufficient coverage to identify these disease neighborhoods, thereby confirming a fundamental hypothesis of interactome-based approaches to human disease [117].

There are, however, subtle differences between the connectivity patterns of functionally related proteins and proteins implicated in the same disease. Genes that jointly perform a biological task are often much more densely connected than genes associated with the same disease [53]. An interpretation of this empirical finding is that *dys*function is typically distributed among several, only loosely connected, functional modules on the interactome. This has important implications for the design of network-based algorithms that aim to identify genes with a certain function or dysfunction. While functional associations may be identified using so-called "community detection" algorithms that target dense node groups [130], the identification of disease-associated genes requires different strategies, as reviewed in Section 9.4.

The relationships among nodes within a certain network neighborhood can be generalized to relationships between neighborhoods. A study of more than 44 000 disease pairs identified a network measure for the interactome-based distance of two disease modules, allowing systematic distinguishing of separated or overlapping disease pairs [117]. While overlapping disease modules correspond to diseases with significant molecular similarities, as well as related symptoms and elevated co-morbidity, diseases whose modules are separated lack any detectable molecular or clinical relationships. These findings were later extended, showing that the network distance of the targets of a particular drug to a disease module is predictive for drug efficacy [54] and can be used for identifying drug repurposing candidates [131, 132].

9.4   Disease Module Analysis

The local aggregation of both physiological and pathobiological processes within interactome networks represents a fundamental biological organization principle that forms the basis for many important applications, ranging from the prediction of protein function to disease gene identification and drug target prioritization. In this section, we will briefly review the process of disease module analysis. A disease module is loosely defined as the comprehensive set of cellular components and their interactions that are associated with a particular disease. Most commonly, a disease module is identified with a connected subgraph within the interactome [3] (Figure 9.3). For most common diseases, such as cardiovascular diseases, cancer or diabetes mellitus, there are hundreds of genes known to be involved. Yet, despite these impressive advances, in particular fueled by sequencing technology, we are still far from a complete understanding of their molecular determinants. For example, more than 2000 genes are estimated to be involved in intellectual disabilities, yet our current knowledge includes only around 800 genes [133]. The basic idea of disease module analysis is to use the connectivity patterns observed among known disease-associated genes to systematically scan their respective interactome neighborhood for genes with a yet unknown important role for the disease. This principle has been applied successfully to a broad range of diseases, from rare Mendelian disorders [134], to cancer [135] and other complex disorders, like metabolic [136], inflammatory [137] or developmental diseases [138].

9.4.1   Seed Cluster Construction

The starting point of the disease module analysis process is the construction of a suitable interactome network (see above for resources) and the curation of known disease-associated genes ("seed genes") for the particular disease of interest. There are

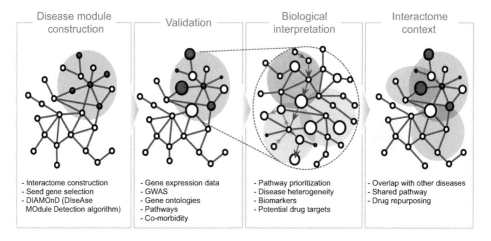

Figure 9.3 Disease module analysis. Overview of the different steps involved in constructing and analyzing the interactome module of a particular disease. The disease module represents all cellular components and their interactions that are responsible for a certain disease. Springer Nature, 2010.

several comprehensive ressources of known disease genes, including OMIM [104], the GWAS catalogue [139] and DisGeNET [140]. The reported disease associations cover a wide spectrum from rare variants with a known and experimentally validated functional mechanism to GWAS variants of rather small effect size and unknown mechanism. Other associations may not be of genetic origin at all, such as differential gene expression or associations inferred solely from text mining. Given this broad variety in possible interaction data and gene–disease associations, a certain trade-off between using only highest-confidence data and achieving the highest possible coverage is unavoidable. We recommend to experiment with different solutions, ideally guided by a domain expert of the specific disease of interest. While there is no simple recipe to tackle this challenging problem of setting up the initial interactome and seed genes, the network perspective can provide some guidance: if the seed cluster is not significantly localized on the network, it is unlikely that a network-based expansion algorithm will be able to identify the relevant neighborhood around this cluster. The connectivity of the initial seed cluster, for example in terms of the statistical significance of the largest connected component size, may therefore provide a rough indication as to whether a particular combination of interactome and seed gene data meets the minimal criteria for a meaningful disease module analysis [141].

9.4.2   Network-Based Disease Gene Prioritization

There are numerous algorithms that scan the network neighborhood around a given set of seed genes. They can be broadly classified into three major categories: (1) connectivity based methods, (2) path-based methods and (3) diffusion-based methods.

Connectivity-Based Methods

These methods build directly on the observation that disease genes tend to interact with each other. Early approaches considered all direct neighbors of the seed cluster as potential candidate genes [142]. More recent refinements of this idea take the degree heterogeneity of the interactome into account [143] or include more advanced relationships among groups of nodes, such as graphlets [144] or connectivity significance [53, 141].

Path-Based Methods

The genes in the neighborhood of the seed cluster can also be ranked according to their network distance to the respective seed genes. There are several possibilities for how to quantify the network distance between sets of nodes, such as using different weighted averages as implemented in [145]. Another option is to search for a set of candidate nodes that collectively minimize the path lengths within the seed cluster, such as using minimum spanning tree (or "Steiner tree") approaches [146–148]. These approaches extend the set of seed genes using a minimal number of additional nodes and edges required for connecting all seed genes in a single connected component.

Diffusion-Based Methods

The closeness of candidate genes to the seed cluster can also be assessed using dynamic approaches, such as diffusion processes [134, 135, 149–152]. A widely used choice is the random walk with restart (RWR) [153]. Starting from the seed genes, a random walker wanders along randomly chosen links of the network until returning to a randomly chosen seed gene (with restart probability r per time step) and starting the process all over. After sufficient iterations of this process, the frequencies with which the individual nodes in the network are visited will converge to a stationary value. This value can then be used to rank the nodes from most related (highest frequency) to least related (lowest frequency) to the seed cluster. The restarting probability r can be used to tune the influence of the seed genes on the diffusive process, from free diffusion (no influence of the seed genes, $r = 0$) to no diffusion at all (walker never leaves the seed genes, $r = 1$).

9.4.3   Validation of the Disease Module

Depending on the chosen prioritization algorithm, the outcome is most commonly given by a ranked list of genes. The next step in the disease module analysis process is to assess the relevance of the identified genes and choose a cutoff, that is, the number of additional candidate genes that will be integrated into the final module. We recommend a complementary strategy that uses cross-validation methods and enrichment with independent biological data.

Cross-Validation of Prediction Performance

Similar to other class prediction tasks, we can use k-fold cross-validation to evaluate the performance of the chosen algorithm. First, the set of original seed genes is randomly divided into k groups. Next, we remove one of the k subgroups from the seed genes and repeat the prioritization procedure using only the remaining $k - 1$ groups as a modified seed gene pool. The performance of the algorithm can then be evaluated from its ability to retrieve the genes from the original seed gene set that was left out. In contrast to many other classification tasks, disease gene prioritization lacks clear true negatives, that is, genes known not to be involved in the disease. Standard performance measures, such as receiver operating characteristic curves, therefore need to be interpreted with some caution. Some approximations have been proposed for true negative genes, such as genes that are essential or unlikely to be involved in a particular disease due to their expression pattern.

Enrichment with Independent Biological Data
A more biologically motivated approach for estimating the relevance of the identified candidate genes is to use independent biological data and test from enrichment. Examples of such complementary data include (1) genes corresponding to GWAS loci of relevant studies, (2) genes found to be differentially expressed in a respective case/control study, (3) genes involved in biological processes or pathways that are known to be relevant to the diseases (expert curated), (4) genes involved in the same pathways or processes as the seed genes (unbiased enrichment) or (5) genes implicated in diseases that show high co-morbidity with the disease under investigation. In [141], for example, the authors used a sliding window approach to assess how the enrichment decreases along the list of ranked candidate genes. This also allows both for evaluating how relevant the candidate genes are, as well as choosing a cutoff rank, beyond which the candidate genes show no strong biological signal.

9.4.4   Interpreting the Disease Module

In the last, and arguably most important, step of the disease module analysis we turn to the biological interpretation of the identified genes and their interactions. We can distinguish two main perspectives, one that focuses on the biological mechanisms contained within the module, and one evaluating the interactome context of the entire module.

Elucidating the Molecular Mechanisms of the Disease
Utilizing the diverse biological data collected at the validation step, we can now extract a more detailed picture of the biological processes that are contained within the disease module. Following a strategy proposed in [141], we first combine the various layers of evidence per gene into a single score. For example, a gene that has been identified in a GWAS study, that is linked to differential expression and that is also known to be involved in a highly co-morbid disease may receive a higher score than a gene with fewer lines of evidence supporting its involvement in the disease. This can be achieved by first ranking all genes separately for each data set and then combining the scores, using for example the so-called Borda count [154] or other methods. Next, we can use the integrated score for prioritizing pathways or biological processes contained in the module, for example through the average score of the contained genes. Alternatively, one can use standard tools for gene set enrichment analyses in [155, 156], or more advanced network-based methods such as in [157, 158].

Context within the Interactome

As discussed above, the network-based relationships between modules may offer biological insights that can also be explored as part of a disease module analysis. Overlaps with other disease modules may reveal mechanisms of frequent co-morbidities, or indicate potential drug repurposing options. A network-based analysis may further identify potential submodules, for example for stratifying patients into subgroups for more personalized treatment.

9.5 Discussion and Outlook

In light of the complex networks of networks that are involved in human disease, this chapter could only offer a first glimpse into the emerging field of network medicine. Important research questions remain for each network introduced above. For most systems, no complete mapping exists yet, on either the node level or the edge level. Parallel to efforts to obtain more and more complete maps, increasingly sophisticated network analysis methods are being developed, hopefully allowing us to gain deeper insights into the complex relationships across the different scales that govern health and disease. Indeed, a major challenge that we are only beginning to address is to go beyond individual networks and include the intricate relationships between them. First steps in this direction consider multi-layer networks, in which different layers represent different relationships among entities or multiplex networks, in which different kinds of interactions are introduced between different types of nodes [159–161]. Such approaches have already been used to study the spread of epidemics by representing different modalities of contact as different layers [162–164]. The different layers may, for example, represent online social networks, public transit use or flight paths between cities, each characterized by different spreading rates and mechanisms. These promising first results indicate the potential that such approaches may hold also for integrating molecular and cellular networks.

References

[1] Greene JA, Loscalzo J. Putting the patient back together-social medicine, network medicine, and the limits of reductionism. *New Eng J Med*. 2017;377(25):2493.

[2] Beadle GW, Tatum EL. Genetic control of biochemical reactions in *Neurospora*. *PNAS*. 1941;27(11):499–506.

[3] Barabási A-L, Gulbahce N, Loscalzo J. Network medicine: a network-based approach to human disease. *Nat Rev Genet*. 2011;12(1):56–68.

[4] McGillivray P, Clarke D, Meyerson W, et al. Network analysis as a grand unifier in biomedical data science. *Ann Rev Biomed Data Sci*. 2018;1:153–180.

[5] Lazebnik Y. Can a biologist fix a radio? Or, what I learned while studying apoptosis. *Cancer Cell*. 2002;2(3):179–182.

[6] Vidal M, Cusick ME, Barabási A-L. Interactome networks and human disease. *Cell*. 2011;144(6):986–998.

[7] Swainston N, Smallbone K, Hefzi H, et al. Recon 2.2: from reconstruction to model of human metabolism. *Metabolomics*. 2016;12:109.

[8] Ma H, Sorokin A, Mazein A, et al. The Edinburgh human metabolic network reconstruction and its functional analysis. *Mol Syst Biol*. 2007;3:135.

[9] Kanehisa M, Furumichi M, Tanabe M, Sato Y, Morishima K. KEGG: new perspectives on genomes, pathways, diseases and drugs. *Nucleic Acids Res*. 2017;45(D1):D353–D361.

[10] Gille C, Bölling C, Hoppe A, et al. HepatoNet1: a comprehensive metabolic reconstruction of the human hepatocyte for the analysis of liver physiology. *Mol Syst Biol*. 2010;6:411.

[11] Zelezniak A, Pers TH, Soares S, Patti ME, Patil KR. Metabolic network topology reveals transcriptional regulatory signatures of type 2 diabetes. *PLoS Comput Biol*. 2010;6(4):e1000729.

[12] Folger O, Jerby L, Frezza C, et al. Predicting selective drug targets in cancer through metabolic networks. *Mol Syst Biol*. 2011;7: 501.

[13] Breitkreutz D, Hlatky L, Rietman E, Tuszynski JA. Molecular signaling network complexity is correlated with cancer patient survivability. *PNAS*. 2012;109(23):9209–9212.

[14] Hecker M, Lambeck S, Toepfer S, et al. Gene regulatory network inference: data integration in dynamic models – a review. *Biosystems*. 2009;96(1): 86–103.

[15] Lee TI, Rinaldi NJ, Robert F, et al. Transcriptional regulatory networks in *Saccharomyces cerevisiae*. *Science*. 2002;298(5594):799–804.

[16] Mathelier A, Fornes O, Arenillas DJ, et al. JASPAR 2016: a major expansion and update of the open-access database of transcription factor binding profiles. *Nucleic Acids Res*. 2016;44(D1):D110–D115.

[17] Wingender E, Dietze P, Karas H, Knüppel R. TRANSFAC: a database on transcription factors and their DNA binding sites. *Nucleic Acids Res*. 1996;24(1): 238–241.

[18] Huntzinger E, Izaurralde E. Gene silencing by microRNAs: contributions of translational repression and mRNA decay. *Nat Rev Genet*. 2011;12(2):99–110.

[19] Filipowicz W, Bhattacharyya SN, Sonenberg N. Mechanisms of post-transcriptional regulation by microRNAs: are the answers in sight? *Nat Rev Genet*. 2008;9(2):102–114.

[20] Ardekani AM, Naeini MM. The role of microRNAs in human diseases. *Avicenna J Med Biotechnol*. 2010;2(4):161–179.

[21] Lu M, Zhang Q, Deng M, et al. An analysis of human microRNA and disease associations. *PLoS One*. 2008;3(10):e3420.

[22] Heyn J, Hinske LC, Ledderose C, Limbeck E, Kreth S. Experimental miRNA target validation. *Methods Mol Biol*. 2013;936:83–90.

[23] Kuhn DE, Martin MM, Feldman DS, et al. Experimental validation of miRNA targets. *Methods*. 2008;44(1):47–54.

[24] Rehmsmeier M, Steffen P, Hochsmann M, Giegerich R. Fast and effective prediction of microRNA/target duplexes. *RNA*. 2004;10(10):1507–1517.

[25] Agarwal V, Bell GW, Nam J-W, Bartel DP. Predicting effective microRNA target sites in mammalian mRNAs. *Elife*. 2015;4.

[26] Krek A, Grün D, Poy MN, et al. Combinatorial microRNA target predictions. *Nat Genet*. 2005;37(5):495–500.

[27] Betel D, Wilson M, Gabow A, Marks DS, Sander C. The microRNA.org resource: targets and expression. *Nucleic Acids Res*. 2008;36(Database issue):D149–D153.

[28] Dweep H, Sticht C, Pandey P, Gretz N. miRWalk-database: prediction of possible miRNA binding sites by "walking" the genes of three genomes. *J Biomed Inform*. 2011;44(5):839–847.

[29] Griffiths-Jones S, Saini HK, van Dongen S, Enright AJ. miRBase: tools for microRNA genomics. *Nucleic Acids Res*. 2008;36(Database issue):D154–D158.

[30] Sethupathy P, Corda B, Hatzigeorgiou AG. TarBase: a comprehensive database of experimentally supported animal microRNA targets. *RNA*. 2006;12(2): 192–197.

[31] Xiao F, Zuo Z, Cai G, et al. mirecords: an integrated resource for microRNA–target interactions. *Nucleic Acids Res*. 2009;37(suppl 1):D105–D110.

[32] Huttlin EL, Ting L, Bruckner RJ, et al. The BioPlex network: a systematic exploration of the human interactome. *Cell*. 2015;162(2):425–440.

[33] Huttlin EL, Bruckner RJ, Paulo JA, et al. Architecture of the human interactome defines protein communities and disease networks. *Nature*. 2017;545(7655): 505–509.

[34] Rolland T, Tasan M, Charloteaux B, et al. A proteome-scale map of the human interactome network. *Cell*. 2014;159(5):1212–1226.

[35] De Las Rivas J, Fontanillo C. Protein–protein interactions essentials: key concepts to building and analyzing interactome networks. *PLoS Comput Biol*. 2010;6 (6):e1000807.

[36] Ofran Y, Rost B. Predicted protein–protein interaction sites from local sequence information. *FEBS Lett*. 2003;544(1–3):236–239.

[37] Gallet X, Charloteaux B, Thomas A, Brasseur A. A fast method to predict protein interaction sites from sequences. *J Mol Biol*. 2000;302:917–926.

[38] Yan C, Dobbs D, Honavar V. A two-stage classifier for identification of protein–protein interface residues. *Bioinformatics*. 2004;20(suppl 1):i371–i378.

[39] Deng M, Mehta S, Sun F, Chen T. Inferring domain–domain interactions from protein–protein interactions. *Genome Res*. 2002;12(10):1540–1548.

[40] Marcotte CJV, Marcotte EM. Predicting functional linkages from gene fusions with confidence. *Appl Bioinformatics*. 2002;1(2):93–100.

[41] Pellegrini M, Marcotte EM, Thompson MJ, Eisenberg D, Yeates TO. Assigning protein functions by comparative genome analysis: protein phylogenetic profiles. *PNAS*. 1999;96(8):4285–4288.

[42] Hakes L, Pinney JW, Robertson DL, Lovell SC. Protein–protein interaction networks and biology – what's the connection? *Nat Biotechnol.* 2008;26(1): 69–72.

[43] Gillis J, Ballouz S, Pavlidis P. Bias tradeoffs in the creation and analysis of protein–protein interaction networks. *J Proteomics.* 2014;100:44–54.

[44] Kerrien S, Aranda B, Breuza L, et al. The IntAct molecular interaction database in 2012. *Nucleic Acids Res.* 2012;40(Database issue):D841–D846.

[45] Hermjakob H, Montecchi-Palazzi L, Lewington C, et al. IntAct: an open source molecular interaction database. *Nucleic Acids Res.* 2004;32(Database issue):D452–D455.

[46] Croft D, Mundo AF, Haw R, et al. The reactome pathway knowledgebase. *Nucleic Acids Res.* 2014;42(Database issue):D472–D477.

[47] Fabregat A, Jupe S, Matthews L, et al. The reactome pathway knowledgebase. *Nucleic Acids Res.* 2018;46(D1):D649–D655.

[48] Alanis-Lobato G, Andrade-Navarro MA, Schaefer MH. HIPPIE v2.0: enhancing meaningfulness and reliability of protein–protein interaction networks. *Nucleic Acids Res.* 2017;45(D1):D408–D414.

[49] Szklarczyk D, Morris JH, Cook H, et al. The STRING database in 2017: quality-controlled protein–protein association networks, made broadly accessible. *Nucleic Acids Res.* 2017;45(D1):D362–D368.

[50] Jeong H, Mason SP, Barabási A-L, Oltvai ZN. Lethality and centrality in protein networks. *Nature.* 2001;411(6833):41–42.

[51] Lim J, Hao T, Shaw C, et al. A protein–protein interaction network for human inherited ataxias and disorders of Purkinje cell degeneration. *Cell.* 2006;125(4):801–814.

[52] Vanunu O, Magger O, Ruppin E, Shlomi T, Sharan R. Associating genes and protein complexes with disease via network propagation. *PLoS Comput Biol.* 2010;6(1):e1000641.

[53] Ghiassian SD, Menche J, Barabási A-L. A DIseAse MOdule detection (DIAMOnD) algorithm derived from a systematic analysis of connectivity patterns of disease proteins in the human interactome. *PLoS Comput Biol.* 2015;11(4):e1004120.

[54] Guney E, Menche J, Vidal M, Barábasi A-L. Network-based in silico drug efficacy screening. *Nat Commun.* 2016;7:10331.

[55] White JG, Southgate E, Thomson JN, Brenner S. The structure of the nervous system of the nematode *Caenorhabditis elegans. Philos Trans R Soc Lond B Biol Sci.* 1986;314(1165):1–340.

[56] Briggman KL, Helmstaedter M, Denk W. Wiring specificity in the direction-selectivity circuit of the retina. *Nature.* 2011;471(7337):183.

[57] Bock DD, Lee W-CA, AM Kerlin, et al. Network anatomy and in vivo physiology of visual cortical neurons. *Nature.* 2011;471(7337):177.

[58] Glasser MF, Coalson TS, Robinson EC, et al. A multi-modal parcellation of human cerebral cortex. *Nature.* 2016;536(7615):171–178.

[59] Sporns O. The human connectome: origins and challenges. *Neuroimage.* 2013;80:53–61.

[60] Yan G, Vértes PE, Towlson EK, et al. Network control principles predict neuron function in the *Caenorhabditis elegans* connectome. *Nature.* 2017;550(7677):519.

[61] Seidlitz J, Váša F, Shinn M, et al. Morphometric similarity networks detect microscale cortical organization and predict inter-individual cognitive variation. *Neuron.* 2018;97(1):231–247.

[62] Xia CH, Ma Z, Ciric R, et al. Linked dimensions of psychopathology and connectivity in functional brain networks. *Nat Commun.* 2018;9(1):3003.

[63] Bergthaler A, Menche J. The immune system as a social network. *Nat Immunol.* 2017;18(5):481.

[64] Rieckmann JC, Geiger R, Hornburg D, et al. Social network architecture of human immune cells unveiled by quantitative proteomics. *Nat Immunol.* 2017;18(5):583.

[65] Kveler K, Starosvetsky E, Ziv-Kenet A, et al. Immune-centric network of cytokines and cells in disease context identified by computational mining of PubMed. *Nature Biotechnol.* 2018;36(7):651–659.

[66] Bernoulli D. Essai dune nouvelle analyse de la mortalité causée par la petite vérole et des avantages de linoculation pour la prévenir. *Histoire de lAcad Roy Sci(Paris) avec Mém des Math et Phys and Mém.* 1760;1:1–45.

[67] Pastor-Satorras R, Castellano C, Van P Mieghem, Vespignani A. Epidemic processes in complex networks. *Rev Mod Phys.* 2015;87(3):925–979.

[68] Longini IM, Jr, Nizam A, Xu S, et al. Containing pandemic influenza at the source. *Science.* 2005;309(5737):1083–1087.

[69] Granell C, Gómez S, Arenas A. Dynamical interplay between awareness and epidemic spreading in multiplex networks. *Phys Rev Lett.* 2013;111(12):128701.

[70] Christakis NA, Fowler JH. The spread of obesity in a large social network over 32 years. *N Engl J Med.* 2007;357(4):370–379.

[71] Christakis NA, Fowler JH. The collective dynamics of smoking in a large social network. *N Engl J Med.* 2008;358(21):2249–2258.

[72] Fowler JH, Christakis NA. Dynamic spread of happiness in a large social network: longitudinal analysis over 20 years in the Framingham Heart Study. *BMJ.* 2008;337:a2338.

[73] Zhang B, Horvath S. A general framework for weighted gene co-expression network analysis. *Stat Appl Genet Mol Biol.* 2005;4:Article 17.

[74] De Smet R, Marchal K. Advantages and limitations of current network inference methods. *Nat Rev Microbiol.* 2010;8(10):717–729.

[75] GTEx Consortium. Human genomics: the Genotype-Tissue expression (GTEx) pilot analysis – multitissue gene regulation in humans. *Science.* 2015;348(6235):648–660.

[76] Weirauch MT. Gene coexpression networks for the analysis of DNA microarray data. In *Applied Statistics for Network Biology.* Wein: Wiley-VCH Verlag, 2011, pp. 215–250.

[77] Parikshak NN, Swarup V, Belgard TG, et al. Genome-wide changes in lncRNA, splicing, and regional gene expression patterns in autism. *Nature*. 2016; 540:423.

[78] Zhang B, Gaiteri C, Bodea L-G, et al. Integrated systems approach identifies genetic nodes and networks in late-onset Alzheimer's disease. *Cell*. 2013;153(3):707–720.

[79] Peters LA, Perrigoue J, Mortha A, et al. A functional genomics predictive network model identifies regulators of inflammatory bowel disease. *Nat Genet*. 2017;49:1437.

[80] Zhang S, Liu W, Liu X, Qi J, Deng C. Biomarkers identification for acute myocardial infarction detection via weighted gene co-expression network analysis. *Medicine*. 2017;96(47):e8375.

[81] Zhang J, Xiang Y, Ding L, et al. Using gene co-expression network analysis to predict biomarkers for chronic lymphocytic leukemia. *BMC Bioinformatics*. 2010;11(suppl. 9): S5.

[82] Wang L-X, Li Y, Chen G-Z. Network-based co-expression analysis for exploring the potential diagnostic biomarkers of metastatic melanoma. *PLoS One*. 2018;13(1):e0190447.

[83] Costanzo M, VanderSluis B, Koch EN, et al. A global genetic interaction network maps a wiring diagram of cellular function. *Science*. 2016;353(6306).

[84] Szappanos B, Kovács K, Szamecz B, et al. An integrated approach to characterize genetic interaction networks in yeast metabolism. *Nat Genet*. 2011;43 (7):656–662.

[85] Tong AHY, Lesage G, Bader GD, et al. Global mapping of the yeast genetic interaction network. *Science*. 2004;303(5659):808–813.

[86] Blomen VA, Májek P, Jae LT, et al. Gene essentiality and synthetic lethality in haploid human cells. *Science*. 2015;350 (6264):1092–1096.

[87] Wang T, Yu H, Hughes NW, et al. Gene essentiality profiling reveals gene networks and synthetic lethal interactions with oncogenic ras. *Cell*. 2017;168(5):890–903.e15.

[88] Srivas R, Shen JP, Yang CC, et al. A network of conserved synthetic lethal interactions for exploration of precision cancer therapy. *Mol Cell*. 2016;63(3):514–525.

[89] Han K, Jeng EE, Hess GT, et al. Synergistic drug combinations for cancer identified in a CRISPR screen for pairwise genetic interactions. *Nat Biotechnol*. 2017;35(5):463.

[90] Moder M, Velimezi G, Owusu M, et al. Parallel genome-wide screens identify synthetic viable interactions between the BLM helicase complex and Fanconi anemia. *Nat Commun*. 2017;8(1):1238.

[91] Kim D, Rossi J. RNAi mechanisms and applications. *Biotechniques*. 2008;44(5): 613–616.

[92] Doench JG, Fusi N, Sullender M, et al. Optimized sgRNA design to maximize activity and minimize off-target effects of CRISPR-Cas9. *Nat Biotechnol.* 2016;34(2):184–191.

[93] Kubicek S, Gilbert JC, Fomina-adlin DY, et al. Chromatin-targeting small molecules cause class-specific transcriptional changes in pancreatic endocrine cells. *PNAS.* 2012;109(14):5364–5369.

[94] Bansal M, Yang J, Karan C, et al. A community computational challenge to predict the activity of pairs of compounds. *Nat Biotechnol.* 2014;32(12):1213–1222.

[95] Markt P, Dürnberger G, Colinge J, Kubicek S. CLOUD: CeMM library of unique drugs. *J Cheminform.* 2012;4(Suppl 1):P23.

[96] Isik Z, Baldow C, Cannistraci CV, Schroeder M. Drug target prioritization by perturbed gene expression and network information. *Sci Rep.* 2015;5:17417.

[97] Bray M-A, Singh S, Han H, et al. Cell painting, a high-content image-based assay for morphological profiling using multiplexed fluorescent dyes. *Nat Protocol.* 2016;11(9):1757.

[98] Zhang F, Gao B, Xu L, et al. Allele-specific behavior of molecular networks: understanding small-molecule drug response in yeast. *PLoS One.* 2013;8(1):e53581.

[99] Noh H, Shoemaker JE, Gunawan R. Network perturbation analysis of gene transcriptional profiles reveals protein targets and mechanism of action of drugs and influenza A viral infection. *Nucleic Acids Res.* 2018;46(6):e34.

[100] Schubert M, Klinger B, Münemann Kl, et al. Perturbation-response genes reveal signaling footprints in cancer gene expression. *Nat Commun.* 2018;9(1):20.

[101] Dorel M, Klinger B, Sieber A, et al. Modelling signalling networks from perturbation data. *bioRxiv*, 2018.

[102] Molinelli EJ, Korkut A, Wang W, et al. Perturbation biology: inferring signaling networks in cellular systems. *PLoS Comput Biol.* 2013;9(12):e1003290.

[103] Goh K-I, Cusick ME, Valle D, et al. The human disease network. *PNAS.* 2007;104(21):8685–8690.

[104] Amberger JS, Bocchini CA, Schiettecatte F, Scott AF, Hamosh A. OMIM.org: Online Mendelian Inheritance in Man (OMIM®), an online catalog of human genes and genetic disorders. *Nucleic Acids Res.* 2015;43(Database issue):D789–D798.

[105] Lee D-S, Park J, Kay K, et al. The implications of human metabolic network topology for disease comorbidity. *PNAS.* 2008;105(29):9880–9885.

[106] Zhou X, Menche J, Barabási A-L, Sharma A. Human symptoms–disease network. *Nat Commun.* 2014;5.

[107] NIH. Medical subject headings. www.nlm.nih.gov/mesh.

[108] Barrenas F, Chavali S, Holme P, Mobini R, Benson M. Network properties of complex human disease genes identified through genome-wide association studies. *PLoS One*. 2009;4(11):e8090.

[109] Zhang M, Zhu C, Jacomy A, Lu LJ, Jegga AG. The orphan disease networks. *Am J Hum Genet*. 2011;88(6):755–766.

[110] Piñero J, Berenstein A, Gonzalez-Perez A, Chernomoretz A, Furlong LI. Uncovering disease mechanisms through network biology in the era of next generation sequencing. *Sci Rep*. 2016;6:24570.

[111] Hidalgo CA, Blumm N, Barabási A-L, Christakis NA. A dynamic network approach for the study of human phenotypes. *PLoS Comput Biol*. 2009;5(4):e1000353.

[112] Chmiel A, Klimek P, Thurner S. Spreading of diseases through comorbidity networks across life and gender. *New J Phys*. 2014;16(11):115013.

[113] Hu JX, Thomas CE, Brunak S. Network biology concepts in complex disease comorbidities. *Nat Rev Genet*. 2016;8.

[114] Duran-Frigola M, Rossell D, Aloy P. A chemo-centric view of human health and disease. *Nat Commun*. 2014;5:5676.

[115] Gomez-Cabrero D, Menche J, Vargas C, et al. From comorbidities of chronic obstructive pulmonary disease to identification of shared molecular mechanisms by data integration. *BMC Bioinformat*. 2016;17(Suppl. 15):441.

[116] Klimek P, Aichberger S, Thurner S. Disentangling genetic and environmental risk factors for individual diseases from multiplex comorbidity networks. *Sci Rep*. 2016;6:39658.

[117] Menche J, Sharma A, Kitsak M, et al. Uncovering disease–disease relationships through the incomplete interactome. *Science*. 2015;347(6224):1257601.

[118] Watts DJ, Strogatz SH. Collective dynamics of "small-world" networks. *Nature*. 1998;393(6684):440–442.

[119] Cohen R, Havlin S. Scale-free networks are ultrasmall. *Phys Rev Lett*. 2003;90(5):058701.

[120] Callaway DS, Newman ME, Strogatz SH, Watts DJ. Network robustness and fragility: percolation on random graphs. *Phys Rev Lett*. 2000;85(25): 5468.

[121] Newman ME, Strogatz SH, Watts DJ. Random graphs with arbitrary degree distributions and their applications. *Phys Rev E*. 2001;64(2):026118.

[122] Cohen R, Erez K, Ben-Avraham D, Havlin S. Resilience of the internet to random breakdowns. *Phys Rev Lett*. 2000;85(21):4626.

[123] Dorogovtsev SN, Mendes JF. *Evolution of Networks: From Biological Nets to the Internet and WWW* . Oxford: Oxford University Press, 2003.

[124] Albert R, Jeong H, Barabási A-L. Error and attack tolerance of complex networks. *Nature*. 2000;406(6794):378–382.

[125] Blomen VA, Májek P, Jae LT, et al. Gene essentiality and synthetic lethality in haploid human cells. *Science*. 2015;350(6264):1092–1096.

[126] Spirin V, Mirny LA. Protein complexes and functional modules in molecular networks. *PNAS*. 2003;100(21):12123–12128.

[127] Hartwell LH, Hopfield JJ, Leibler S, Murray AW. From molecular to modular cell biology. *Nature*. 1999;402(6761 Suppl.):C47–C52.

[128] Barabási A-L, Oltvai ZN. Network biology: understanding the cell's functional organization. *Nat Rev Genet*. 2004;5(2):101–113.

[129] Feldman I, Rzhetsky A, Vitkup D. Network properties of genes harboring inherited disease mutations. *PNAS*. 2008;105(11):4323–4328.

[130] Fortunato S. Community detection in graphs. *Phys Rep*. 2010;486(3–5):75–174.

[131] Langhauser F, Casas AI, Guney E, et al. A diseasome cluster-based drug repurposing of soluble guanylate cyclase activators from smooth muscle relaxation to direct neuroprotection. *NPJ Syst Biol Appl*. 2018;4(1):8.

[132] Cheng F, Desai RJ, Handy DE, et al. Network-based approach to prediction and population-based validation of in silico drug repurposing. *Nat Commun*. 2018;9(1):2691.

[133] Vissers LELM, Gilissen C, Veltman JA. Genetic studies in intellectual disability and related disorders. *Nat Rev Genet*. 2016;17(1):9–18.

[134] Smedley D, Köhler S, Czeschik JC, et al. Walking the interactome for candidate prioritization in exome sequencing studies of Mendelian diseases. *Bioinformatics*. 2014;30(22):3215–3222.

[135] Leiserson MDM, Vandin F, Wu H-T, et al. Pan-cancer network analysis identifies combinations of rare somatic mutations across pathways and protein complexes. *Nat Genet*. 2015;47(2):106–114.

[136] Chen Y, Zhu J, Lum PY, et al. Variations in DNA elucidate molecular networks that cause disease. *Nature*. 2008;452(7186):429–435.

[137] Peters LA, Perrigoue J, Mortha A, et al. A functional genomics predictive network model identifies regulators of inflammatory bowel disease. *Nat Genet*. 2017;49(10):1437–1449.

[138] Krishnan A, Zhang R, Yao V, et al. Genome-wide prediction and functional characterization of the genetic basis of autism spectrum disorder. *Nat Neurosci*. 2016;19(11):1454–1462.

[139] MacArthur J, Bowler E, Cerezo M, et al. The new NHGRI-EBI catalog of published genome-wide association studies (GWAS catalog). *Nucleic Acids Res*. 2016;45(D1):D896–D901.

[140] Piñero J, Bravo A, Queralt-Rosinach N, et al. Disgenet: a comprehensive platform integrating information on human disease-associated genes and variants. *Nucleic Acids Res*. 2016;gkw943.

[141] Sharma A, Menche J, Huang CC, et al. A disease module in the interactome explains disease heterogeneity, drug response and captures novel pathways and genes in asthma. *Hum Mol Genet*. 2015;24(11):3005–3020.

[142] Oti M, Snel B, Huynen MA, Brunner HG. Predicting disease genes using protein–protein interactions. *J Med Genet*. 2006;43(8):691–698.

[143] Erten S, Bebek G, Ewing RM, Koyut Mürk, et al. DADA: degree-aware algorithms for network-based disease gene prioritization. *BioData Min.* 2011;4(1).

[144] Wang X-D, Huang J-L, Yang L, et al. Identification of human disease genes from interactome network using graphlet interaction. *PLoS One.* 2014;9(1):e86142.

[145] Guney E, Oliva B. Exploiting protein–protein interaction networks for genome-wide disease–gene prioritization. *PLoS One.* 2012;7(9):e43557.

[146] Bailly-Bechet M, Borgs C, Braunstein A, et al. Finding undetected protein associations in cell signaling by belief propagation. *PNAS.* 2011;108(2):882–887.

[147] Tuncbag N, McCallum S, Huang S-SC, Fraenkel E. SteinerNet: a web server for integrating "omic" data to discover hidden components of response pathways. *Nucleic Acids Res.* 2012;40(Web Server issue):W505–W509.

[148] Tuncbag N, Gosline SJC, Kedaigle A, et al. Network-based interpretation of diverse high-throughput datasets through the omics integrator software package. *PLoS Comput Biol.* 2016;12(4):e1004879.

[149] Krauthammer M, Kaufmann CA, Gilliam TC, Rzhetsky A. Molecular triangulation: bridging linkage and molecular-network information for identifying candidate genes in Alzheimer's disease. *PNAS.* 2004;101(42):15148–15153.

[150] Vanunu O, Magger O, Ruppin E, Shlomi T, Sharan R. Associating genes and protein complexes with disease via network propagation. *PLoS Comput Biol.* 2010;6(1):e1000641.

[151] Vandin F, Upfal E, Raphael BJ. Algorithms for detecting significantly mutated pathways in cancer. *J Comput Biol.* 2011;18(3):507–522.

[152] Cowen L, Ideker T, Raphael BJ, Sharan R. Network propagation: a universal amplifier of genetic associations. *Nat Rev Genet.* 2017;18:551–562.

[153] Köhler S, Bauer S, Horn D, Robinson PN. Walking the interactome for prioritization of candidate disease genes. *Am J Hum Genet.* 2008;82(4):949–958.

[154] Van Erp M, Schomaker L. Variants of the Borda count method for combining ranked classifier hypotheses. In *Seventh International Workshop on Frontiers in Handwriting Recognition*, 2000.

[155] Subramanian A, Tamayo P, Mootha VK, et al. Gene set enrichment analysis: a knowledge-based approach for interpreting genome-wide expression profiles. *PNAS.* 2005;102(43):15545–15550.

[156] Kuleshov MV, Jones MR, Rouillard AD, et al. Enrichr: a comprehensive gene set enrichment analysis web server 2016 update. *Nucleic Acids Res.* 2016;44(W1):W90–W97.

[157] Merico D, Isserlin R, Stueker O, Emili A, Bader GD. Enrichment map: a network-based method for gene-set enrichment visualization and interpretation. *PLoS One.* 2010;5(11):e13984.

[158] Glaab E, Baudot A, Krasnogor N, Schneider R, Valencia A. Enrichnet: network-based gene set enrichment analysis. *Bioinformatics*. 2012;28(18):i451–i457.

[159] Cozzo E, Baños RA, Meloni S, Moreno Y. Contact-based social contagion in multiplex networks. *Phys Rev E*. 2013;88(5):050801.

[160] Boccaletti S, Bianconi G, Criado R, et al. The structure and dynamics of multilayer networks. *Phys Rep*. 2014;544(1):1–122.

[161] Kivelä M, Arenas A, Barthelemy M, et al. Multilayer networks. *J Complex Netw*. 2014;2(3):203–271.

[162] Min Y, Hu J, Wang W, et al. Diversity of multilayer networks and its impact on collaborating epidemics. *Phys Rev E*. 2014;90(6):062803.

[163] Chen X, Wang R, Tang M, et al. Suppressing epidemic spreading in multiplex networks with social-support. *New J Phys*. 2018;20(1):013007.

[164] Guo Q, Jiang X, Lei Y, et al. Two-stage effects of awareness cascade on epidemic spreading in multiplex networks. *Phys Rev E*. 2015;91(1):012822.

10 Towards a Multi-layer Network Analysis of Disease: Challenges and Opportunities Through the Lens of Multiple Sclerosis

Jesper Tegnér, Ingrid Kockum, Mika Gustafsson and David Gomez-Cabrero

10.1 Introduction

The book you are holding in your hand contains several chapters on networks of networks from different viewpoints. In this chapter, we ask what relevance, if any, could a networks-of-networks approach have in the context of diseases? In particular, would such a formulation be useful from a practical and conceptual standpoint? From a clinical and biological perspective, we like to find robust and reliable biomarkers for diagnosis, prognosis and response to therapy [1, 2]. This is indeed a practical quest. In addition, we would also like to query the biological mechanisms of such biomarkers and their putative contribution to the disease itself. If successful, it would contribute to a deeper understanding of disease and be of conceptual value. At the first level of analysis, we may *evidently* say that since there is such a large abundance of molecular data already published, and even more so in production, we need to organize this information in a systematic way [3]. Furthermore, since such different molecular entities are not independent, there is a need to relate them to each other and consider them as joint data [4, 5]. Hence, it is therefore very attractive to consider a networks-of-networks formulation to be a preferred choice of representing such data and their inherent relations [6, 7] – thus, in this sense, using networks of networks as a bookkeeping tool. Nevertheless, such a perspective masks the conceptual and algorithmic challenge on how to *practically* link different molecular data types to each other. Third, even if we somehow succeed in doing so, we may well ask how useful such an approach would be in a disease context. It may well turn out that such an ambitious endeavour is a dead-end due to sparsity of data and the intricacies of how to integrate different data types [8].

In this chapter we will, therefore, address the current state of affairs on such an approach in a disease context using multiple sclerosis (MS) as an example [9, 10]. In particular, we use MS to survey part of the available molecular data types (DNA [11], mRNA [12, 13], miRNA [14], DNA methylation [15–17] and chromatin accessibility [18]) and to provide three examples of multi-omics studies targeting MS. We close the chapter with an outlook on outstanding challenges for a multi-layer networks-of-network analysis of MS.

Let us begin with a brief account of MS and how it is traditionally analysed from a genetic (DNA) standpoint. MS is a chronic inflammatory disease of the central nervous system (CNS). The disease results in demyelination and neuronal loss due to the autoimmune destruction of myelin and subsequent loss of neurons. In most cases, MS begins as a relapsing–remitting (RR) disease characterized by episodes of clinical symptoms followed by partial or complete recovery. Part of MS patients, approximately 60% after 40 years of disease [19], develop a secondary progressive (SP) disease course, accompanied by a continuous decline in neurological function. Unfortunately, MS is a leading cause of neurological disability in young adults. The rationale for what is referred to as a traditional DNA-based analysis is that we know that genetics play a significant role in this disease, with heritability estimated to be 64% [20] . International consortium studies have identified 233 different genetic variants across the genome, contributing to the pathogenesis of MS [11]. Importantly, the inheritance of the disease is consistent with the different loci only exerting a modest effect with odds ratios (ORs) below 1.4. The only exception thus far is the human leukocyte antigen (HLA) loci, where several polymorphisms affect the risk of MS and the major risk allele *DRB1*15:01* has an OR of 3. The exact cause of MS, however, remains unknown. The inheritance pattern of the disease suggests that one locus exerts a moderate effect whereas the other loci only contribute with modest effects.

Interestingly, the current compendium of MS variants provides a glimpse into other immune-related diseases, such as psoriasis (0.9% overlap) and inflammatory bowel disease, Crohn's disease and primary biliary cirrhosis (all 9.1% overlap). This data is suggestive of a network of molecules affected by such genetic variants and that there are commonalities across diseases as well as disease-specific network structures [21]. In summary, the genetic markers can be viewed as solid data pointing to putative mechanisms involved in the genesis (and possibly progression) of the disease. Yet, the effect sizes of the genetic variants are as a rule tiny, which is a general challenge in DNA-based genetic research. For example, why are the effect sizes small even for a phenotype such as height, despite the validated strong genetic component? This and other examples have generated vivid debate over the years on how to find the *'missing heritability of complex diseases'*. It remains a bottleneck, and we still lack appropriate computational tools to fully capitalize on existing GWAS data, as the analysis is limited by the assumption that SNPs are independent. Here, we could anticipate that a network-based analysis taking into account interaction effects between genetic variants will open up new possibilities for deeper understanding.

Finally, it is known that lifestyle exposures such as smoking, body mass index, sun exposure and viral infections affect the susceptibility to MS [22]. As a result, there is a considerable body of work aiming to disentangle the interactions between such environmental factors and genetic variants.

Recent technological advances have opened new windows into the cellular circuitry beyond the DNA sequence and individual SNPs as detected by GWAS. For example, encyclopaedic analysis of genomes reveals a collection of molecular entities such as DNA methylation, protein-coding RNA, noncoding RNA, splice variants, RNA editing, histone modifications, transcription factors (TFs), transcription start sites, promoters, accessible regions, localization of proteins, several different types of protein modifications and metabolic profiling [23]. As a rule, all of these aspects can be altered in disease. To find a representational language to capture how these different elements are working together is one motivation for a networks-of-networks analysis of these systems.

Here we would like to emphasize the abundance of different molecular entities and interactions, such as those listed above, which occurs between the level of DNA and the environment. This is very much one motivation for a networks-of-networks formulation and analysis for diseases, where we are keen to disentangle mechanisms underlying biomarkers and their interactive effects. Yet, this is, not surprisingly, easier said than done.

To review such challenges, and considering MS as a case study, in the first section we outline (part of) the kinds of data we have between DNA and the environmental data which have been shown to be of relevance in MS. Next, we review some of the known interactions among those layers. Finally, we discuss the use of networks of networks as a tool.

10.2   Relevant Molecular Layers Mediating Between DNA and the Environment

MicroRNAs (miRNAs) are small non-coding RNAs (ncRNAs), usually 20–25 nucleotides long, and were first discovered in 1993. It was observed that lin-4 in *C. elegans* does not code for a protein, but instead produced a pair of short RNA transcripts that each regulate the timing of larval development by translational repression of lin-14, which encodes for a nuclear protein [24, 25]. Since then, it has been established that miRNA regulates gene expression by translational repression, mRNA cleavage, and mRNA decay, which have been found to control cell division, differentiation and death. Such regulation, referred to as regulation occurring at a post-transcriptional level, acts by miRNA binding to the coding region as well as 3' and 5' untranslated regions (UTRs) of messenger RNAs (mRNAs). Such bindings result in either inhibition of translation or degradation of mRNAs [26]. Approximately 30% of the protein-coding genes are influenced by this type of regulation exerted by the ∼1500 miRNAs encoded in the human genome. Notably, an individual miRNA can target multiple genes, and each protein-coding gene can be regulated by several miRNAs, affecting, for example, apoptosis, proliferation, differentiation and metastasis [27, 28]. Since the discovery of small noncoding RNAs, it has become abundantly clear that this was part of a much broader phenomenon in which the genome is literally full of noncoding transcripts, whereas only 1–2% of the transcripts code for proteins. As miRNA displays different expression profiles from tissue to tissue, reflecting the diversity in cellular phenotypes, it is not surprising that it cannot be neglected when analysing diseases. For example, it has been shown in expression correlation studies that all intronic miRNAs are co-transcribed along with their host gene. In MS, changes in miRNA expression in immune cells has been associated with the disease [14, 29, 30].

DNA methylation results from the addition of a methyl group to cytosine residues in the DNA to form 5-methylcytosine (5-mC), and in mammals it is predominantly restricted to the context of CpG dinucleotides [31]. In the human genome, 70–80% of CpG sequences are methylated. CpG islands (CGI) are sequences with high $C + G$ content that are generally unmethylated and colocalize with more than half of the promoters of human genes. Housekeeping genes generally contain a CGI in the neighbourhood of their TSS (transcription start site). Methylation of, for example, CpG shores, genomic regions up to 2 kb distant from CGI, show lower CpG density but increased variability in DNA methylation. Functionally, DNA methylation at enhancer elements can influence the binding of transcription factors (TFs) [31]. Moreover, the genotype is a strong source of inter-individual variability in DNA methylation. Interestingly, growing evidence shows that several environmental factors can influence DNA methylation. Examples include cigarette smoking [32] and air pollution [33], which modifies the epigenetics. Physical exercise can also operate as a physiological stimulus, which can cause changes in DNA methylation [34]. In summary, the relationship between DNA methylation and transcriptional activation/repression is complex. For example, it remains unsolved how this complex interplay is regulated and whether DNA methylation changes are a consequence of TF binding or whether they drive enhancer activity through the exclusion of TF [31]. DNA methylation has been shown to be modified in multiple cell types in MS, including immune cells [13, 17] and brain cells [35], among others. Furthermore, the link between smoking, MS and DNA methylation has been investigated [32].

Histone modifications. Histones are key players because, through covalent modifications of their residuals (e.g. lysine), they have a crucial role in the regulation of transcription, DNA repair and replication [36, 37]. Modifications such as methylation, sumoylation, ubiquitination and acetylation are also dynamically regulated by chromatin-modifying enzymes. Notably, such enzymes are available docking sites in histones and then recruit additional chromatin modifiers and remodelling enzymes. Enzymes are associated with specific histone modifications [38, 39]. There is emerging evidence associating histone modifications and risk for MS [40].

Chromatin accessibility. While histone modifications and chromatin accessibility are mechanistically related, technologies such as DNase-seq [41] and ATAC-seq [42] have allowed the profiling of genome-wide chromatin accessibility on an unprecedented scale. Briefly, the ATAC-seq technique allows the identification of accessible DNA regions by using the hyperactive mutant Tn5 transposase, which inserts sequencing adapters (a process named '*tagmentation*') into open regions of the genome. The *tagged* DNA fragments can then be profiled by NGS technologies [42]. Identification, profiling and quantification of chromatin open regions provides information on which transcription factors (TFs) [43] may have access to regulatory regions such as enhancers and promoters [44]. The characterization of chromatin accessibility in MS is an open area of study, but a very necessary one considering the continuous identification of novel TFs associated with MS, such as EBNA2 [45], NF-κB [46], STAT3 [46], Sox10 [47] and Nurr1 [48], among others.

10.3   Integrating Several Layers

The next logical step in the analysis is addressing how we could combine two or more of these data types in the analysis of MS; how this can be done from a technical standpoint; and if such an integrative analysis would be informative for understanding of MS, such as discovery of biomarkers and prognostic features, and elucidation of the mechanisms driving MS. Here, we illustrate this road ahead by using three recent example studies which share the aim of injecting molecular features and mechanisms between the DNA string and the phenotype of disease itself.

MS through the lens of expression quantitative trait loci. In the genome there is a widespread presence of expression quantitative trait loci, referred to as eQTLs [49, 50]. A large proportion of the eQTLs appears to be active in specific cell types and in response to molecular signalling induced by intrinsic or extrinsic stimuli. Since the majority of non-HLA MS single-nucleotide polymorphisms (SNPs) are noncoding, it is likely that a number of MS-associated loci would be involved in regulation of gene expression, acting as eQTLs [11]. This would provide a gateway for gene expression alterations to impact biological pathways involved in disease. In a recent study, the investigators (some of them co-authors in this chapter) analysed MS susceptibility variants as eQTLs in their disease and cell-type specific context [51]. To this end, whole transcriptome RNA sequencing (RNA-seq) was performed in peripheral blood mononuclear cells (PBMCs) from MS patients who had previously been genotyped on the ImmunoChip SNP microarray. Furthermore, the significance with respect to cell types of the identified MS eQTLs was assessed in lymphoblastic cell lines (LCLs) and primary immune cells (CD4$^+$, CD8$^+$, B cells and monocytes). This analysis also provided functional evidence for most relevant eQTLs by mapping to enhancer histone marks and predicted transcription factor binding sites. This study established that 40% of the eQTLs were found to be more prominent as compared to a non-inflammatory control group of neurological disease patients. Additional eQTL-derived insights are provided in the most resent MS GWAS publication [11].

MS through the lens of DNA methylation. Epigenetics, the other layers of networks and regulatory mechanisms, beyond the pure DNA sequence, can mediate the functional downstream effect of a genetic variant. For example, a recent study reported that DNA methylation could mediate a protective effect through hypomethylation of the HLA-DRB1*15:01 gene, which is a major risk factor for MS [16]. The investigators used a technique referred to as causal inference and Mendelian randomization to provide evidence that *HLA* variants mediate risk for MS via changes in the *HLA-DRB1* DMR that modify *HLA-DRB1* expression. The rationale behind the study was that despite the increasing list of MS-associated genetic variants and regions, it remains challenging to understand how they contribute to the disease; hence epigenetic markers, such as DNA methylation, may provide a mechanistic understanding.

MS through the lens of gene regulatory networks. The fact that disease-associated SNPs are highly enriched for eQTLs in MS and other complex diseases suggests a synergistic upstream interaction in the disease process through the dysregulation of key upstream TFs, which acts through gene regulatory networks (GRNs [52]). A problem or a challenge in the identification of GRNs is that they are highly cell-specific [53], and need to be reverse-engineered, a process that is hampered by a general sparsity of relevant time-series multi-omics data [54]. Thus the analysed GRNs are generally incomplete, and as TFs may have a significant impact on cell differentiation [55] we expect only subtle changes in TF activity in patients, making identification difficult. One successful approach to identify key MS-associated TFs constructed GRNs of human Th1/Th2 differentiations using predicted DNA bindings and mRNA time-series data [56]. Topological network analysis jointly with the positioning of MS-associated SNPs predicted that the master Th2 regulator GATA3 would be upstream-dysregulated in MS. Next, analysis of MS patients in relapse showed that GATA3 target genes were generally differentially expressed, and that one of its isoform splice variants was found to be differentially expressed during remission, which suggests it might be an upstream regulator of relapses. Another approach to characterize MS disease dynamics in CD4+ T-cells was performed by Hellberg et al. [57], who used dynamical mRNA differential expression to identify an MS module of 81 proteins that colocalized in the protein interaction network. Interestingly, these proteins were also highly enriched for harbouring disease-associated SNPs, and measurement of its secreted products revealed new prognostic markers that accurately predicted disease severity two years later. In summary, these examples exemplify the principle of how enrichment of disease associations within networks of networks can be performed using cross-validation [58].

10.4   Networks of Networks: A Necessary Tool

In closing the chapter we ask, given the multitude of different molecular layers and mechanisms between the DNA string and a disease readout, what do we need in order to enable a full-scale multi-layer networks-of-networks analysis of disease. Second, what challenges and advantages can we expect from such an undertaking? One may challenge the idea of the necessity of a networks-of-networks analysis. Clearly, it is possible to find predictive patterns from a single data type in relation to a phenotypic readout [59]. The analysis of SNPs for MS serves as one demonstration of the feasibility of detecting the occurrence of, or an increased risk of, MS. However, as soon as – we would like to argue – one would like to uncover mechanisms of disease, and to develop a personalized precision medicine approach, we need as comprehensive a picture as possible of the intricate involvement of the different molecular layers in different kinds of cells and tissues [60]. Furthermore, from a theoretical standpoint, it is clear that the network dynamics is different when considering interactions between different types of networks as compared to a single network. Thus, there are several reasons supporting the idea that a networks-of-networks analysis of diseases is both necessary and points to the future of systems and precision medicine. If successful, the advantages are huge in that such a project would provide us with a deep under-standing of health and disease from a personalized molecular medicine perspective. Yet, the challenges are significant, both from a data-collection point-of-view, as well as when considering the technical algorithmic challenges on how to actually integrate such data to the effect of being able to disentangle such networks-of-networks struc-tures. Here we need, as a community, to work on how to integrate sparse data. For example, we can expect few patient samples, partial sampling of a subset of data types in each patient, data types in part overlapping between patients and development of algorithms enabling networks-of-networks inferences [61]. We anticipate exciting development of data-driven techniques, incorporating a merger between careful bioinformatics and machine-learning algorithms to address the technical challenges [62]. From a clinical standpoint, it is mandatory that such models of networks of networks are interpretable in case they are to be used for clinical decisions, thus making pure black-box machine-learning approaches less attractive.

10.5   Conclusions

We find that the future of medicine cannot avoid embracing biological complexity in order to enable a development of a personalized precision medicine paradigm. The road ahead, in our view, will most likely be to deal with the molecular complexity operating between the environment (disease) and DNA, in a cell- and tissue-specific manner. Here, we have indicated how and why a networks-of-networks formulation provides an appropriate language to address this challenge for precision medicine.

References

[1] Auffray C, Charron D, Hood L. Predictive, preventive, personalized and participatory medicine: back to the future. *Genome Med*. 2010;2:57. doi:10.1186/gm178

[2] Hood L, Friend SH. Predictive, personalized, preventive, participatory (P4) cancer medicine. *Nat Rev Clin Oncol*.2011;8:184–187. doi:10.1038/nrclinonc .2010.227

[3] Langmead B, Nellore A. Cloud computing for genomic data analysis and collaboration. *Nat Rev Genet*.2018. doi:10.1038/nrg.2018.8

[4] Vidal M, Cusick ME, Barabási A-L. Interactome networks and human disease. *Cell*. 2011;144:986–998. doi:10.1016/j.cell.2011.02.016

[5] Menche J, Sharma A, Kitsak M, et al. Uncovering disease–disease relationships through the human interactome. *Science*. 2015;347. doi:10.1126/science.1257601

[6] Mangioni G, Jurman G, De Domenico M. Multilayer flows in molecular networks identify biological modules in the human proteome. *IEEE Trans Netw Sci Eng*. 2018. doi:10.1109/TNSE.2018.2871726

[7] De Domenico M, Solé-Ribalta A, et al. Mathematical formulation of multilayer networks. *Phys Rev X*. 2014;3:1–15. doi:10.1103/PhysRevX.3.041022

[8] Roca J, Cano I, Gomez-Cabrero D, Tegnér J. From systems understanding to personalized medicine: lessons and recommendations based on a multidisciplinary and translational analysis of COPD. *Meth Mol Biol*. 2016. doi:10.1007/978-1-4939-3283-2

[9] Kingwell K. Disease mechanisms in MS: informing tactics to combat MS. *Nat Rev Neurol*. 2012;8:589. doi:10.1038/nrneurol.2012.218

[10] Ascherio A, Munger KL, Lünemann JD. The initiation and prevention of multiple sclerosis. *Nat Rev Neurol.* 2012;8:602–612. doi:10.1038/nrneurol.2012.198

[11] Patsopoulos NA, Baranzini SE, Santaniello A, et al. Multiple sclerosis genomic map implicates peripheral immune cells and microglia in susceptibility. *Science.* 2019;365:143933. doi:10.1126/science.aav7188

[12] Wood H. Disease mechanisms in MS: RNA profiling uncovers two distinct subsets of patients with multiple sclerosis. *Nat Rev Neurol.* 2012;8:591. doi:10.1038/nrneurol.2012.217

[13] Fernandes SJ, Morikawa H, Ewing E, et al. Non-parametric combination analysis of multiple data types enables detection of novel regulatory mechanisms in T cells of multiple sclerosis patients. *Sci Rep.* 2019;9:11996. doi:10.1038/s41598-019-48493-7

[14] Ruhrmann S, Ewing E, Piket E, et al. Hypermethylation of MIR21 in CD4+ T cells from patients with relapsing–remitting multiple sclerosis associates with lower miRNA-21 levels and concomitant up-regulation of its target genes. *Mult Scler J.* 2018;24:1288–1300. doi:10.1177/1352458517721356

[15] Zeitelhofer M, Adzemovic MZ, Gomez-Cabrero D, et al. Functional genomics analysis of vitamin D effects on CD4+ T cells in vivo in experimental autoimmune encephalomyelitis. *PNAS.* 2017;114. doi:10.1073/pnas.1615783114

[16] Kular L, Liu Y, Ruhrmann S, et al. DNA methylation as a mediator of HLA-DRB1 15:01 and a protective variant in multiple sclerosis. *Nat Commun.* 2018;9. doi:10.1038/s41467-018-04732-5

[17] Ewing E, Kular L, Fernandes SJ, et al. Combining evidence from four immune cell types identifies DNA methylation patterns that implicate functionally distinct pathways during multiple sclerosis progression. *Ebiomedicine.* 2019;43:411–423. doi:10.1016/j.ebiom.2019.04.042

[18] Elias S, Schmidt A, Gomez-Cabrero D, Tegner J. Gene regulatory network of human GM-CSF secreting T helper cells. *bioRxiv.* doi:10.1101/555433

[19] Fambiatos A, Jokubaitis V, Horakova D, et al. Risk of secondary progressive multiple sclerosis: a longitudinal study. *Mult Scler J.* 2019. doi:10.1177/1352458519868990

[20] Westerlind H, Ramanujam R, Uvehag D, et al. Modest familial risks for multiple sclerosis: a registry-based study of the population of Sweden. *Brain.* 2014;137:770–778. doi:10.1093/brain/awt356

[21] Olafsson S, Stridh P, Bos SD, et al. Fourteen sequence variants that associate with multiple sclerosis discovered by meta-analysis informed by genetic correlations. *Genomic Med.* 2017;2:24. doi:10.1038/s41525-017-0027-2

[22] Olsson T, Barcellos LF, Alfredsson L. Interactions between genetic, lifestyle and environmental risk factors for multiple sclerosis. *Nat Rev Neurol.* 2016;13:25. doi:10.1038/nrneurol.2016.187

[23] Metzker ML. Sequencing technologies: the next generation. *Nat Rev Genet.* 2010;11:31–46. doi:10.1038/nrg2626

[24] Feinbaum R, Ambros V, Lee R. The *C. elegans* heterochronic gene lin-4 encodes small RNAs with antisense complementarity to lin-14. *Cell.* 2004;116:843–854.

[25] Wightman B, Ha I, Ruvkun G. Posttranscriptional regulation of the heterochronic gene lin-14 by lin-4 mediates temporal pattern formation in C. *elegans. Cell.* 1993;75:855–862. doi:10.1016/0092-8674(93)90530-4

[26] Khorshid M, Hausser J, Zavolan M, van Nimwegen E. A biophysical miRNA–mRNA interaction model infers canonical and noncanonical targets. *Nat Methods.* 2013;10:253–255. doi:10.1038/nmeth.2341

[27] Allmer J. The role of MicroRNAs in biological processes. *miRNomics.* 2014;1107:177–187. doi:10.1007/978-1-62703-748-8

[28] Piriyapongsa J, Mariño-Ramírez L, Jordan IK. Origin and evolution of human microRNAs from transposable elements. *Genetics.* 2007;176:1323–1337. doi:10.1534/genetics.107.072553

[29] Gandhi R. miRNA in multiple sclerosis: search for novel biomarkers. *Mult Scler J.* 2015;21:1095–1103. doi:10.1177/1352458515578771

[30] Baulina N, Kulakova O, Kiselev I, et al. Immune-related miRNA expression patterns in peripheral blood mononuclear cells differ in multiple sclerosis relapse and remission. *J Neuroimmunol.* 2018;317:67–76. doi:10.1016/j.jneuroim.2018.01.005

[31] Jones P. Functions of DNA methylation: islands, start sites, gene bodies and beyond. *Nat Rev Genet.* 2012. doi:10.1038/nrg3230

[32] Marabita F, Almgren M, Sjöholm LK, et al. Smoking induces DNA methylation changes in multiple sclerosis patients with exposure–response relationship. *Sci Rep.* 2017;7:14589. doi:10.1038/s41598-017-14788-w

[33] Rider CF, Carlsten C. Air pollution and DNA methylation: effects of exposure in humans. *Clin Epigenetics.* 2019;11:131. doi:10.1186/s13148-019-0713-2

[34] Lindholm ME, Marabita F, Gomez-Cabrero D, et al. An integrative analysis reveals coordinated reprogramming of the epigenome and the transcriptome in human skeletal muscle after training. *Epigenetics.* 2014;9:1557–1569. doi:10.4161/15592294.2014.982445

[35] Kular L, Needhamsen M, Adzemovic MZ, et al. Neuronal methylome reveals CREB-associated neuro-axonal impairment in multiple sclerosis. *Clin Epigenetics.* 2019;11:86. doi:10.1186/s13148-019-0678-1

[36] Kornberg RD. Chromatin structure? A repeating unit of histones and DNA. *Science.* 1974;184:3–6.

[37] Herceg Z, Murr R. Mechanisms of histone modifications. In *Handbook of Epigenetics.* Amsterdam: Elsevier, 2011. doi:10.1016/B978-0-12-375709-8.00003-4

[38] Poleshko A, Kossenkov AV, Shalginskikh N, et al. Human factors and pathways essential for mediating epigenetic gene silencing. *Epigenetics.* 2014;9:1280–1289.

[39] Liang Z, Brown KE, Carroll T, et al. A high-resolution map of transcriptional repression. *Elife*. 2017;6:1–24. doi:10.7554/eLife.22767

[40] He H, Hu Z, Xiao H, Zhou F, Yang B. The tale of histone modifications and its role in multiple sclerosis. *Hum Genomics*. 2018;12:31. doi:10.1186/s40246-018-0163-5

[41] Thurman RE, Rynes E, Humbert R, et al. The accessible chromatin landscape of the human genome. *Nature*. 2012;489:75–82. doi:10.1038/nature11232

[42] Buenrostro JD, Giresi PG, Zaba LC, Chang HY, Greenleaf WJ. Transposition of native chromatin for fast and sensitive epigenomic profiling of open chromatin, DNA-binding proteins and nucleosome position. *Nat Methods*. 2013;10: 1213–1218. doi:10.1038/nmeth.2688

[43] Weirauch MT, Cote A, Norel R, et al. Evaluation of methods for modeling transcription factor sequence specificity. *Nat Biotechnol*. 2013;31:126–134. doi:10.1038/nbt.2486

[44] Hoffman MM, Ernst J, Wilder SP, et al. Integrative annotation of chromatin elements from ENCODE data. *Nucleic Acids Res*. 2013;41:827–841. doi:10.1093/nar/gks1284

[45] Ricigliano VAG, Handel AE, Sandve GK, et al. EBNA2 binds to genomic intervals associated with multiple sclerosis and overlaps with vitamin d receptor occupancy. *PLoS One*. 2015;10:e0119605. doi:10.1371/journal .pone.0119605

[46] Nuzziello N, Vilardo L, Pelucchi P, et al. Investigating the role of microRNA and transcription factor co-regulatory networks in multiple sclerosis pathogenesis. *Int J Mol Sci*. 2018. doi:10.3390/ijms19113652

[47] Mokhtarzadeh Khanghahi A, Satarian L, Deng W, Baharvand H, Javan M. In vivo conversion of astrocytes into oligodendrocyte lineage cells with transcription factor Sox10: Promise for myelin repair in multiple sclerosis. *PLoS One*. 2018;13:e0203785. doi:10.1371/journal.pone.0203785

[48] Montarolo F, Martire S, Perga S, Bertolotto A. NURR1 impairment in multiple sclerosis. *Int J Mol Sci*. 2019. doi:10.3390/ijms20194858

[49] Sun W, Hu Y. eQTL mapping using RNA-seq data. *Stat Biosci*. 2012. doi:10.1007/s12561-012-9068-3

[50] Gilad Y, Rifkin SA, Pritchard JK. Revealing the architecture of gene regulation: the promise of eQTL studies. *Trends Genet*. 2008. doi:10.1016/j.tig.2008.06.001

[51] James T, Linde M, Morikawa H, et al. Impact of genetic risk loci for multiple sclerosis on expression of proximal genes in patients. *Hum Mol Genet*. 2018. doi:10.1093/hmg/ddy001

[52] Gustafsson M, Hörnquist M, Lundström J, Björkegren J, Tegnér J. Reverse engineering of gene networks with LASSO and nonlinear basis functions. *Ann N Y Acad Sci*. 2009;1158:265–275. doi:10.1111/j.1749-6632.2008.03764.x

[53] Marbach D, Lamparter D, Quon G, et al. Tissue-specific regulatory circuits reveal variable modular perturbations across complex diseases. *Nat Methods*. 2016;13:366–370. doi:10.1038/nmeth.3799

[54] Tan K, Tegnér J, Ravasi T. Integrated approaches to uncovering transcription regulatory networks in mammalian cells. *Genomics*. 2008;91:219–231. doi:https://doi.org/10.1016/j.ygeno.2007.11.005

[55] Rackham OJL, Firas J, Fang H, et al. A predictive computational framework for direct reprogramming between human cell types. *Nat Genet*. 2016;48:331. doi:10.1038/ng.3487

[56] Gustafsson M, Gawel DR, Alfredsson L, et al. A validated gene regulatory network and GWAS identifies early regulators of T cell–associated diseases. *Sci Transl Med*. 2015;7:313ra178–313ra178. doi:10.1126/scitranslmed.aad2722

[57] Hellberg S, Eklund D, Gawel DR, et al. Dynamic response genes in CD4+ T cells reveal a network of interactive proteins that classifies disease activity in multiple sclerosis. *Cell Rep*. 2016;16:2928–2939. doi:10.1016/j.celrep.2016.08.036

[58] Gustafsson M, Nestor CE, Zhang H, et al. Modules, networks and systems medicine for understanding disease and aiding diagnosis. *Genome Med*. 2014;6:1–11.

[59] Erusalimsky JD, Grillari J, Grune T, et al. In search of 'omics'-based biomarkers to predict risk of frailty and its consequences in older individuals: the FRAILOMIC Initiative. *Gerontology*. 2015;62:182–190. doi:10.1159/000435853

[60] Tegnér JN, Compte A, Auffray C, et al. Computational disease modeling: fact or fiction? *BMC Syst Biol*. 2009;3:56. doi:10.1186/1752-0509-3-56

[61] Hofmann-Apitius M, Ball G, Gebel S, et al. Bioinformatics mining and modeling methods for the identification of disease mechanisms in neurodegenerative disorders. *Int J Mol Sci*. 2015. doi:10.3390/ijms161226148

[62] Gomez-Cabrero D, Tegnér J. Iterative systems biology for medicine: time for advancing from network signatures to mechanistic equations. *Curr Opin Syst Biol*. 2017;3:111–118. doi:10.1016/j.coisb.2017.05.001

11 Microbiome: A Multi-layer Network View is Required

Rodrigo Bacigalupe, Saeed Shoaie and David Gomez-Cabrero

11.1 Introduction: Microbiome, the Next Frontier

11.1.1 What is the Microbiome?

Microorganisms are ubiquitously present in every environment on Earth, including the interiors and surfaces of humans. Communities of microbes living on or within our bodies comprise bacteria, archaea, viruses and unicellular eukaryotes, which are globally known as the human microbiota. Although often used interchangeably, the term microbiome refers to the entire habitat of the microbiota, including the microbes, their genomes and the environment [1].

The microbiota is estimated to be composed of as many bacteria as human cells [2] and plays a key role in the maintenance of the health status of the host. It is established at birth, when the organs of the newborn are colonized by bacteria from the maternal microbiota and the environment. Subsequently, during development, the microbiota undergoes dynamic changes, influenced by feeding practices, therapies, illness, physiological variation within the host and other exogenous or endogenous factors [3]. The microbiota has essential functions for human physiology, the correct development of the host immune system, defence against infections, synthesis of essential compounds and regulation of the energy balance. For example, microorganisms of the gut microbiota produce enzymes that contribute to the degradation of complex substrates that humans cannot otherwise digest [4], while others biosynthesize essential amino acids and vitamins that are not produced by the human host [5]. Similarly, microbes on the skin flora contribute to the development of immunity and maintenance of the epidermal barrier [6]; and the microbiota of the respiratory tract provides resistance to colonization by respiratory pathogens [7] (Figure 11.1).

In recent years, the human microbiome has gained enormous attention due to its association with an increasing number of disorders and diseases, and probably representing a causal factor in several of them, such as obesity, diabetes and certain types of cancer [8]. Therefore, determining the composition, functions and factors

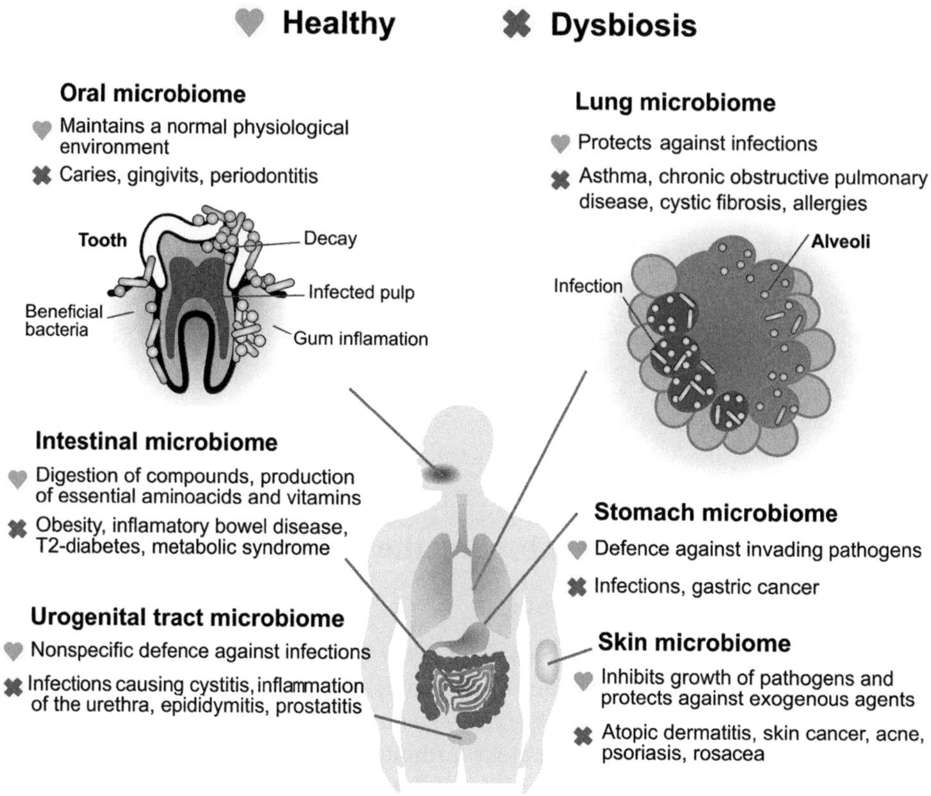

Figure 11.1 The human microbiota in health and disease. Microbes residing in different body locations have a broad range of functions for the healthy development of the immune system and maintenance of the healthy state of individuals. Dysbiosis or alterations of microbial communities have been identified in multiple diseases.

promoting the stability of microbiomes from different body locations is crucial to fully understand the underlying mechanisms responsible for the health-state of individuals. With this in mind, the US and European governments financed two very large-scale projects for characterizing the human microbiota and examining its role in human health and disease: the Human Microbiome Project (HMP) and the MetaHIT initiative, respectively. These projects brought together multiple scientific experts that explored the microbial communities and their relationships with their human hosts [9]. The analysis of samples from hundreds of individuals exhibiting a wide range of health statuses and physiological characteristics allowed them to find relationships between the microbiota and various diseases. In addition, reference catalogues of microbial genome sequences from different body locations, such as the skin, mouth, gastrointestinal tract, urogenital tract or lungs were generated (www.human-microbiome.org).

Identifying the molecular processes associated with specific diseases can provide excellent targets to develop biomarkers for disease detection or therapeutic applications. Recent experiments indicate that microbial communities can be manipulated through dietary modification, antibiotics and probiotics, representing a promising therapeutic avenue to treat disorders caused or exacerbated by imbalances in the microbiota [10].

11.1.2 Recent Technologies in Microbiome and Bioinformatics

With the continuous development of high-throughput technologies and novel bioinformatics techniques, 'multi-*omics*' studies that integrate genetic, functional and metabolic activities of the human microbiota have become feasible [11].

Early human microbiome studies were limited to identifying the taxonomic profiles present in a particular niche through sequencing of the 16S ribosomal RNA (rRNA) gene, a highly conserved, genus or species-specific gene that permits discriminating between different taxonomical groups. This approach requires extraction of bulk DNA from samples, sequencing of PCR products and comparisons of sequences to a reference database. Although several studies have used this method to characterize the composition and structure of microbiomes in health and disease states, it presents some biases and limitations [8]. With next-generation sequencing (NGS) technologies, the entire genomic material of microbial communities can also be obtained. Following DNA extraction from samples and library preparation, shotgun sequencing can be performed with the Illumina HiSeq or similar technologies, producing millions of sequencing reads that have to be analysed using bioinformatics tools. After performing quality control of the reads, these can be mapped to reference genomes or assembled into metagenomes, which facilitate identifying the genetic components of microbial taxa and the construction of microbial gene networks. This approach, known as WGS metagenomics, provides information of the microbiome at a deeper structural level and with great resolution, revealing gene functions found in specific microbes and gene groups harboured by several strains. Nevertheless, such approaches fail to inform on the molecular interactions between the microbial communities and the host, which result from the expression of genes, the interaction between the proteins they encode and the levels of secondary metabolites produced.

The metatranscriptome, consisting of all the RNA molecules expressed by the microbiome, represents the next -*omic* layer and provides a comprehensive understanding of the microbial activity. Metatranscriptomic RNAseq requires extracting the total RNA from the microbiome, followed by cDNA synthesis and sequencing of libraries. Next, reads can be assembled de novo to produce a metatranscriptome or mapped to functional databases, such as KEGG, for revealing the regulation and expression profiles of microbial communities. This database also provides annotations on the molecular interactions, reactions and relation networks for metabolism, genetic information processing or cellular processes, among others. However, a better understanding of the microbiota functions can be achieved through metaproteomics, which allows us to characterize the entire pool of proteins present in the microbiome at any given time. A typical metaproteomic study requires an efficient extraction of proteins from the entire microbial community, followed by the pre-fractionation of peptides. Clean-up and depletion of host cells are essential steps to avoid host-cell contamination. Next, samples are subjected to liquid-chromatography-based separation and mass spectrometry (MS); and the data produced is parsed into bioinformatics pipelines for identification, quantification and functional characterization of proteins [12].

Finally, meta-metabolomics is a powerful tool to analyse the global pool of metabolites present in the microbiome, produced by both the microbiota and the host. Contrary to other -*omic* approaches, the metabolome is the final downstream product and the closest to the phenotype. It is very diverse, more complex, and reflects physiological and clinical biomarkers that cannot be obtained using other methods [13]. Current technologies used for detection, characterization and quantification of metabolites include mass spectroscopy linked to chromatography and nuclear magnetic resonance (NMR), with minimal preparation required for biofluid samples, facilitating assays with high reproducibility [14].

Overall, these technologies can generate enormous amounts of data that require sophisticated bioinformatics tools and statistical analyses. When multiple -*omics* are available, different profiles can be integrated for further data exploration using multivariate statistics, permitting better description of the relationships between the microbes and the host, and correlating alterations with disease-related phenotypes [15].

11.1.3 Microbiome in Health and Disease

The human microbiota, as a result of being deeply embedded within the human body, has a profound impact in the host. Simultaneously, the host controls the microbial ecosystem structure, composition and function to maintain homeostasis. For example, the gut microbiome is controlled by the epithelial cells of the intestine through limitation of available oxygen in the colon. This physiological process promotes the growth of obligate anaerobic bacteria of the Firmicutes and Bacteroidetes phyla and prevents the expansion of facultative aerobic enterobacteria and potentially pathogenic strains [16]. The correct maintenance of this balance ensures the gut microbiome acts as a *microbial* organ that produces metabolites that contribute to host nutrition, development of the immune system and niche protection. When the composition of the microbial community shifts and the gut homeostasis is disrupted, some microbiota-derived metabolites are depleted or generated at harmful concentrations [16], resulting in dysbiotic states that can be associated with multiple disorders, including obesity, inflammatory bowel disease, metabolic syndrome and type 2 diabetes [17]. Similarly, communities of microbes inhabiting the skin are controlled by several types of epidermal glands and skin immunocytes. Being the largest organ, the physical and chemical features of the skin vary enormously across multiple body sites, with different anatomical locations showing diverse appendages and glandular structures [6]. Consequently, microbial communities also differ greatly in their compositions, but overall act as gatekeepers providing resistance to colonization and infection by pathogens and protection against the external environment [7]. Certain skin disorders such as atopic *dermatitis, psoriasis* or *hidradenitis suppurativa* are correlated with alterations of the microbiota, resulting in inflammatory responses that can lead to skin damage [18].

Despite an increasing number of studies describing links between the microbiota composition and specific disorders, the molecular mechanisms used by the microbial communities and the host for affecting each other are still unclear. In addition, it is often unknown whether alterations in the microbiota are the cause of the disease, the result from underlying conditions, or consequence of antibiotics used to treat an infection. For instance, in high-income countries, overuse of antibiotics and changes in diet have led to microbiota that lacks the flexibility and diversity needed for balanced immune responses, probably accounting for some of the dramatic increase of autoimmune and inflammatory disorders [19]. The microbiota has also been related to cancer susceptibility and can modulate the carcinogenic process. For instance, the intestinal microbiota can promote and prevent cancer development and influence therapeutic outcomes through immune responses, and the metabolite produced [20].

These examples highlight the high level of interrelation between the host and the microbiota, with complex and sophisticated microbe–microbe and microbe–host interactions. This complexity can be studied using systems-oriented graph-theoretical approaches, with networks of genes, mRNAs, proteins or metabolites representing the complex interactions between the microbiota and the host [21]. Alterations in the network structure and node connectivity could reflect aberrant situations corresponding to disease states. Despite the utility of network models for capturing and displaying the complexity of microbial interactions, there are several challenges for their interpretation and for translating these analyses into disease management strategies.

11.2   Genome-Scale Metabolic Models

As shown before, the recent advances in multi-omics data generation and ability to link the biological entities and components have led to the emergence of understanding of genotype–phenotype relationships as the fundamental concept of the systems biology approach [22]. Systems biology presents a promising approach for understanding multi-omic high-throughput data and characterizing its properties. The availability of the organism's full genome sequence led the field to generating genome-scale network reconstruction as the predictive platform to connect the specific state of the network to its phenotype. One of the major parts of the genome encodes the metabolic functions. This fact directed the very first network reconstruction on the genome-scale model of cellular metabolism [23]. These models embody a set of biochemical reactions as an output from the gene products within a cell, tissue, organ or organism. Genome-scale metabolic models (GEMs) connect these reactions with each other to give a comprehensive view of cellular metabolism, while linking them to the gene–protein information. GEMs can be represented as a mathematical formulation so they can be understandable for computational tools. The main component behind this mathematical property is stoichiometric information on biochemical reactions in the cell or organism. This knowledge builds up the stoichiometric (S) matrix, which consists of the metabolites and reactions as its components. The S matrix denotes the biochemical connectivity of the GEMs for the target organism. This matrix is fundamental to several applications of the GEMs to quantify the model. The detail of the GEM reconstruction is beyond the scope of this book, and it has been well documented elsewhere [24, 25]. One of the main applications of the GEM is constraint-based modelling, in which there are sets of mathematical constraints imposed to narrow down possible functional cellular states to optimize a specific cellular objective function [26]. Constraint types can be varied from uptake and secretion rates to omics data integration of the target organism in its specific functional state. These constraints are imposed to optimize the cellular functions for specific objective functions, such as biomass production or maximum ATP generation [27]. The optimization method behind the problems differs based on the linearity of the constraints and the objective functions. One of the simple cases for constraint-based modelling is the flux balance analysis, in which both constraints and objective function are linear, and the optimization problem can be solved by linear programming that returns one specific solution to the objective, governing the imposed constraints [25]. One of the other main applications of the GEMs is using the topology and connectivity of these networks to integrate multi-omics data. The omics data can determine the topological enrichment in the network to denote which pathway or region of the network is significantly perturbed in different functional states [28]. As the GEMs connect the metabolites and genes, omics data integration could select candidate metabolites that are significantly affected [29]. To date, GEMs have been used for integrative data analyses and recently for obtaining clinical data and conducting studies on mice, humans and microbiomes [30, 31].

During the past years there have been several studies focused on the single-organism, genome-scale metabolic reconstruction and applications, such as bacteria, eukaryotes and archaea. As reported by Gu et al., 6239 GEMs had been reconstructed as of February 2019 [32]. There is also tremendous work on the reconstruction of the genome-scale model for human metabolism [33], and these efforts even go beyond only the gene–protein reaction and involve the structural modelling of enzymes as well [34].

In the area of microbiomes, GEM reconstruction for the human gut bacteria has received lots of attention. Due to lack of high-throughput methods to elucidate the metabolic interactions between microbes and microbes and microbes and hosts, as well as diet and drugs in the complex microbial ecosystems, GEM reconstruction for the microbial community has received a great deal of consideration following large-scale studies on the human microbiome association with diseases. The initial studies focused on the manual GEM reconstruction of representative species from prevalent microbial phyla in the human gut microbiome [35]. In 2017, human gut bacterial GEMs were reconstructed for 773 members of the human gut biota in a semi-automatic fashion [31]. These GEMs to some extent could explain the bacteria-specific metabolism of the overall human gut microbiome; however, due to the nature of the automatic reconstruction of these GEMs, one needs to perform further curation to have better predictive results [36].

In addition to the individual application of bacterial GEMs in the microbiome, these models can be used in community-level analysis to simulate the metabolic interactions between species while predicting the growth, uptake and secretion of metabolites from each species and their communities. Pairwise interactions between the species by constructing a community S matrix consisting of the species-specific S matrix and an extra shared compartment in the matrix was part of an early attempt to perform community flux balance analysis [37]. Recently, several algorithms for metabolic community modelling at a larger scale have been developed. Among them, optCom introduced multi-level optimization by setting up individual and community objective functions and determined a solution that maximized the individual and systems-level objectives [38]. The multi-level optimization together with the community network interactions as the weights for the objective functions have been used to quantify diet changes in the metabolic interactions in human gut bacteria [30]. These methods could also be used through minimization optimization and constrainting of bacterial growth to predict micronutrient uptake, which could be used to simulate diet changes to improve phenotypes.

As mentioned before, the reconstruction of bacterial GEMs is mainly developed around the full genome sequence of the species. However, in the microbiome studies, there are several unclassified species with just a draft metagenome assembled genome (MAGs)[39]. There are several missing pieces of information in the functional annotation of MAGs that makes the reconstruction of the GEMs for them a challenge. Usually, researchers generate metabolic modules consisting of a set of biochemical reactions for the MAGs without precise connectivity. In the future, GEM science needs to adapt its reconstruction methods to the incomplete or assembled genomes with better ability to fill gaps and to take extra steps to find the interactions between components of the species and community networks.

11.3   Multi-layer Network Approach for Understanding Microbiomes

GEMs are not the only tools to study the microbiome using (possibly multi-)omic information. Interestingly, using different -omics technologies and network approaches, previous studies have investigated microbiomes associated with several diseases from ecological and evolutionary perspectives (dynamics of microbial communities, phylogenetic relationships) and at various molecular levels (gene content, gene expression, protein levels and metabolites interactions). For example, gene co-expression networks correlating transcript levels in different conditions can reveal genes controlled by similar transcriptional factors, being functionally related or belonging to the same protein complexes. Likewise, the complex interactions of metabolites produced by microbial communities create specific functional networks when data from healthy microbiomes is compared to certain disorders. Although these studies have transformed our understanding of microbiota, individual networks ignore the information associated with other biological levels and provide a reductionist overview [40]. A complete understanding of the structure, functions and dynamics of microbial communities at a systems biology level requires analysis at higher dimensional scales.

11.3.1   The Connections Between Layers

We consider that, by integrating individual networks into multi-layer networks, we can represent and quantify the interactions arising from the connections between individual components of the microbiota and the human body, with the potential to reveal novel insights regarding the global mechanisms regulating the functions and processes of microbiomes. For instance, the amount of enzymes that produce metabolites

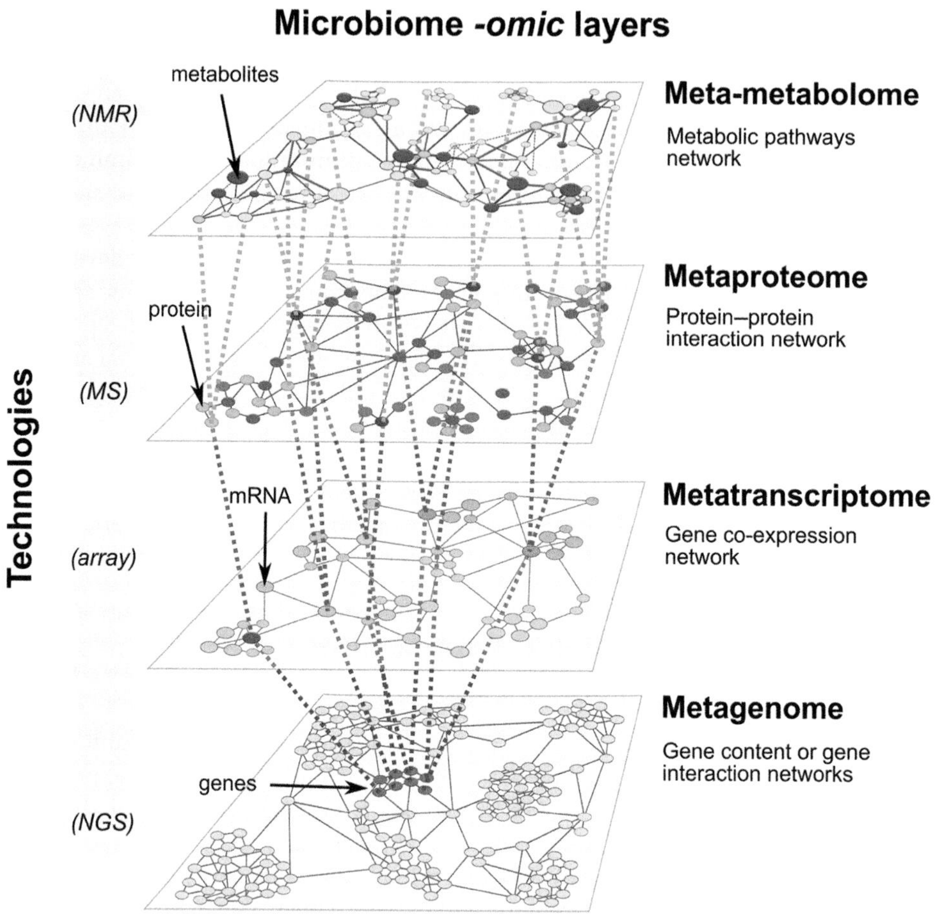

Microbiome -*omic* layers

Figure 11.2   Network of networks of major biological components in microbiomes. Networks of metabolic pathways, PPIs, co-expression and gene content are represented from top to bottom. Interactions between layers originating from a cluster of genes (bottom, blue) are increasingly complex.

can be transcriptionally regulated by transcription factors (TFs) or other global mechanisms. At the same time, the expression levels of genes encoding TFs can be regulated by specific metabolites, which can be produced by other microbiota members or via post-translational modifications by histidine kinases that sense factors produced by the host. As an example, Figure 11.2 depicts four network layers (genes, transcripts, proteins and metabolites) that are dynamic and in continuous communication with each other to determine the metabolic state of individual components in the microbiome.

There are two major approaches for reconstructing networks of multi-omic data. The first method uses prior knowledge of molecular interactions available in public databases, and the multi-layer network provides causality and input–output relationships at the molecular level [41]. The second method for constructing integrated networks is data-driven and infers associations and correlations between molecules using statistical models on multi-omic data. Although this approach does not require prior knowledge of the molecular interactions, the connections between layers do not reflect real biochemical networks.

The next level of connection would be between cell types, where every cell (e.g. microbiome) can have a model of itself interacting with other microbes. The same strategy can be used to model microbiome–host cell–cell interactions [40].

11.3.2 Multi-layer Networks: Opportunities and Limitations for Understanding the Microbiome

Once the existence of several layers and the connections between them are establish, the use of multi-layered networks [42] is a *logical* next step. The use of multiplex networks [43] should be considered when working with layers having the same nodes; this situation could be considered when creating and integrating networks combining information from different organisms, either several microbes or host–microbiome interactions. The use of non-multiple multi-layered networks is expected to be more relevant when considering the integration of, for instance, metabolic information with cell–cell interactions. As a result, we consider that microbiome research can benefit from existing tools and well-established theory on multiplex networks, but it can work as a perfect scenario for real applications of generic multi-layered networks.

References

[1] Marchesi JR, Ravel J. The vocabulary of microbiome research: a proposal. *Microbiome.* 2015;3:31. doi:10.1186/s40168-015-0094-5

[2] Sender R, Fuchs S, Milo R. Revised estimates for the number of human and bacteria cells in the body. *PLoS Biol.* 2016. doi:10.1371/journal.pbio.1002533

[3] D'Argenio V, Salvatore F. The role of the gut microbiome in the healthy adult status. *Clin Chim Acta.* 2015;451:97–102. doi:10.1016/j.cca. 2015.01.003

[4] Flint HJ, Scott KP, Louis P, Duncan SH. The role of the gut microbiota in nutrition and health. *Nat Rev Gastroenterol Hepatol.* 2012;9:577–589. doi:10.1038/nrgastro.2012.156

[5] Qin J, Li R, Raes J, et al. A human gut microbial gene catalogue established by metagenomic sequencing. *Nature.* 2010;464:59–65. doi:10.1038/nature08821

[6] Sanford JA, Gallo RL. Functions of the skin microbiota in health and disease. *Semin Immunol.* 2013;25:370–377. doi:10.1016/j.smim.2013.09.005

[7] Man WH, de Steenhuijsen Piters WAA, Bogaert D. The microbiota of the respiratory tract: gatekeeper to respiratory health. *Nat Rev Microbiol.* 2017;15:259.

[8] Grice EA, Segre JA. The human microbiome: our second genome. *Annu Rev Genomics Hum Genet.* 2012;13:151–170. doi:10.1146/annurev-genom-090711-163814

[9] Methé BA, Nelson KE, Pop M, et al. A framework for human microbiome research. *Nature.* 2012;486:215–221. doi:10.1038/nature11209

[10] Foxx-Orenstein AE. New and emerging therapies for the treatment of irritable bowel syndrome: an update for gastroenterologists. *Therap Adv Gastroenterol.* 2016;9:354–375. doi:10.1177/1756283X16633050

[11] Peterson J, Garges S, Giovanni M, et al. The NIH Human Microbiome Project. *Genome Res.* 2009;19:2317–2323.

[12] Petriz BA, Franco OL. Metaproteomics as a complementary approach to gut microbiota in health and disease. *Front Chem.* 2017. doi:10.3389/fchem.2017 .00004

[13] Chadeau-Hyam M, Ebbels TMD, Brown IJ, et al. Metabolic profiling and the metabolome-wide association study: significance level for biomarker identification. *J Proteome Res.* 2010;9:4620–4627. doi:10.1021/pr1003449

[14] Vernocchi P, Del Chierico F, Putignani L. Gut microbiota profiling: metabolomics based approach to unravel compounds affecting human health. *Front Microbiol.* 2016;7:1144. doi:10.3389/fmicb.2016.01144

[15] Del Chierico F, Gnani D, Vernocchi P, et al. Meta-omic platforms to assist in the understanding of NAFLD gut microbiota alterations: tools and applications. *Int J Mol Sci.* 2014. doi:10.3390/ijms15010684

[16] Byndloss MX, Bäumler AJ. The germ–organ theory of non-communicable diseases. *Nat Rev Microbiol.* 2018;16:103. doi:10.1038/nrmicro.2017.158

[17] Barko PC, McMichael MA, Swanson KS, Williams DA. The gastrointestinal microbiome: a review. *J Vet Intern Med.* 2018;32:9–25. doi:10.1111/jvim.14875

[18] Chen YE, Fischbach MA, Belkaid Y. Skin microbiota–host interactions. *Nature.* 2018;553:427. doi:10.1038/nature25177

[19] Belkaid Y, Hand TW. Role of the microbiota in immunity and inflammation. *Cell.* 2014;157:121–141. doi:10.1016/j.cell.2014.03.011

[20] Pope JL, Tomkovich S, Yang Y, Jobin C. Microbiota as a mediator of cancer progression and therapy. *Transl Res.* 2017;179:139–154. doi:10. 1016/j.trsl .2016.07.021

[21] Witherden EA, Moyes DL, Bruce KD, Ehrlich SD, Shoaie S. Using systems biology approaches to elucidate cause and effect in host–microbiome interactions. *Curr Opin Syst Biol.* 2017;3:141–146. doi:10.1016/ j.coisb.2017.05.003

[22] Gomez-Cabrero D, Tegnér J. Iterative systems biology for medicine: time for advancing from network signatures to mechanistic equations. *Curr Opin Syst Biol.* 2017;3. doi:10.1016/j.coisb.2017.05.001

[23] Edwards JS, Palsson BO. Systems properties of the *Haemophilus influenzae* Rd metabolic genotype. *J Biol Chem.* 1999;274:17410–17416. doi:10.1074/jbc.274.25.17410

[24] Thiele I, Palsson BO. A protocol for generating a high-quality genome-scale metabolic reconstruction. *Nat Protoc.* 2010;5:93–121. doi:10.1038/ nprot.2009.203

[25] Orth JD, Thiele I, Palsson BO. What is flux balance analysis? *Nat Biotechnol.* 2010;28:245–248. doi:10.1038/nbt.1614

[26] Bordbar A, Monk JM, King ZA, Palsson BO. Constraint-based models predict metabolic and associated cellular functions. *Nat Rev Genet.* 2014;15:107–120. doi:10.1038/nrg3643

[27] Schuetz R, Kuepfer L, Sauer U. Systematic evaluation of objective functions for predicting intracellular fluxes in *Escherichia coli. Mol Syst Biol.* 2007;3:119. doi:10.1038/msb4100162

[28] Oberhardt MA, Palsson BO, Papin JA. Applications of genome-scale metabolic reconstructions. *Mol Syst Biol.* 2009;5:320. doi:10.1038/msb.2009.77

[29] Patil KR, Nielsen J. Uncovering transcriptional regulation of metabolism by using metabolic network topology. *PNAS.* 2005;102:2685–2689. doi:10.1073/pnas.0406811102

[30] Shoaie S, Ghaffari P, Kovatcheva-Datchary P, et al. Quantifying diet-induced metabolic changes of the human gut microbiome. *Cell Metab.* 2015;22:320–331. doi:10.1016/j.cmet.2015.07.001

[31] Magnusdottir S, Heinken A, Kutt L, et al. Generation of genome-scale metabolic reconstructions for 773 members of the human gut microbiota. *Nat Biotechnol.* 2017;35:81–89. doi:10.1038/nbt.3703

[32] Gu C, Kim GB, Kim WJ, Kim HU, Lee SY. Current status and applications of genome-scale metabolic models. *Genome Biol.* 2019;20:121. doi:10.1186/s13059-019-1730-3

[33] Thiele I, Swainston N, Fleming RMT, et al. A community-driven global reconstruction of human metabolism. *Nat Biotechnol.* 2013;31:1–15. doi:10.1038/nbt.2488

[34] Brunk E, Sahoo S, Zielinski DC, et al. Recon3D enables a three-dimensional view of gene variation in human metabolism. *Nat Biotechnol.* 2018;36:272–281. doi:10.1038/nbt.4072

[35] Shoaie S, Karlsson F, Mardinoglu A, et al. Understanding the interactions between bacteria in the human gut through metabolic modeling. *Sci Rep.* 2013;3:2532. doi:10.1038/srep02532

[36] Babaei P, Shoaie S, Ji B, Nielsen J. Challenges in modeling the human gut microbiome. *Nat Biotechnol.* 2018;36:682–686. doi:10.1038/nbt.4213

[37] Baldini F, Heinken A, Heirendt L, et al. The microbiome modeling toolbox: from microbial interactions to personalized microbial communities. *Bioinformatics.* 2019;35:2332–2334. doi:10.1093/bioinformatics/bty941

[38] Zomorrodi AR, Maranas CD. OptCom: a multi-level optimization framework for the metabolic modeling and analysis of microbial communities. *PLoS Comput Biol.* 2012;8:e1002363. doi:10.1371/journal.pcbi.1002363

[39] Nayfach S, Shi ZJ, Seshadri R, Pollard KS, Kyrpides NC. New insights from uncultivated genomes of the global human gut microbiome. *Nature.* 2019;568:505–510. doi:10.1038/s41586-019-1058-x

[40] Layeghifard M, Hwang DM, Guttman DS. Disentangling Interactions in the microbiome: a network perspective. *Trends Microbiol.* 2017;25:217–228. doi:10.1016/j.tim.2016.11.008

[41] Yugi K, Kubota H, Hatano A, Kuroda S. Trans-omics: how to reconstruct biochemical networks across multiple 'omic' layers. *Trends Biotechnol.* 2016;34:276–290. doi:10.1016/j.tibtech.2015.12.013

[42] De Domenico M, Solé-Ribalta A, Cozzo E, et al. Mathematical formulation of multilayer networks. *Phys Rev X.* 2014;3:1–15. doi:10.1103/PhysRevX.3.041022

[43] Nicosia V, Bianconi G, Latora V, Barthelemy M. Growing multiplex networks. *Phys Rev Lett.* 2013;111:1–5. doi:10.1103/PhysRevLett.111.058701

PART V
CONCLUSION

12 Concluding Remarks: Open Questions and Challenges

*Ginestra Bianconi, David Gomez-Cabrero, Jesper Tegnér
and Narsis A. Kiani*

12.1 Introduction

During the last decade, we have witnessed an acceleration in the production of data, commonly referred to as big data. This has in turn created a need for tools capable of not only analysing such data sets, but also providing an efficient representation of the data. It is well known that an efficient representation sets the stage for an effective and informative analysis. For example, in the analysis of signals it makes perfect sense to use a Fourier basis if the signals have a smooth behaviour, but the analysis requires many more coefficients to represent the data if the primary signals are discontinuous. The current production of big data lends itself naturally to a network representation of the data, in which nodes could represent observables and the interactions – computed or observed – are represented by edges. We can therefore conclude that networks are here to stay and there exists a body of mature work enabling analysis of networks across numerous fields of science, ranging from the natural sciences to the social sciences. The crux of the matter – as discussed in the present book – is that when multiple connected yet different data types are produced, we end up with what we have referred to as networks of networks, in essence nodes and edges of different semantics. Here in the concluding chapter we will summarize some of the take-home messages and insights from the chapters and outline pertinent open questions and challenges that we believe will take centre-stage in the next five years.

12.2 Major Insights on Multi-layer Networks

In the last 20 years the single-molecule framework has shown its limitations, and system biology has extensively used complex networks to represent the rich web of interactions in cells and to predict cell function. However, until now most of the research has focused on single networks – that is networks formed by molecular interactions of the same nature forming, for instance, transcription networks, metabolic networks, protein–protein interaction networks and signalling networks. Recently it has been realized that complex biological systems are rarely formed by a single network. Rather, they are typically constituted by several interacting networks that cannot be captured by monoplexes. Ignoring this aspect of systems, existing subsystems, can yield misleading and apparently contradictory results.

Here, we advocate that in order to face the current challenges in molecular biology it is necessary to embrace further the complexity of the cell and combine different sets of interactions and sources of data using the multi-layer network framework.

Multi-layer networks or networks of networks allow the representation, prediction and inference of data sets in which the interactions can have different connotations [1]. While multi-layer networks can have an arbitrary topology, it is often useful to focus on a restricted multi-layer network topology, the multiplex network, that has been widely investigated in network theory in the last five years.

Multiplex networks are formed by different layers. Each layer is a network formed by the same number of nodes, where nodes in different layers are in one-to-one correspondence. Multiplex networks have a structure that is much richer than monoplexes. Therefore, from multiplex networks it is possible to infer much more information than from their single layers studied in isolation. Moreover, multi-layer networks and multiplex network can sustain dynamical processes that have no equivalent in single networks, as can be shown by comparing the multi-layer network dynamics to the dynamics projected to the aggregated network. In the last ten years [1] the field of multi-layer networks and multiplex networks has been rapidly expanding, providing a rich set of tools to tackle the complexity of biological systems, and currently it constitutes an ideal framework to uncover the interplay between the architecture of living systems and their function.

However, the field is still young and important challenges will need to be addressed in the coming years to allow this framework to become the central tool in system biology and personalized medicine.

These are:

1. development of multi-layer system biology tools;
2. a comprehensive approach for multi-layer network inference;
3. further development of software for analysis and visualization of multi-layer networks; and
4. progress on dynamics on multi-layer networks.

In the following we discuss these challenges with the hope that this book will help to further stimulate research in these directions.

12.2.1   Developments of Multi-layer Biological Networks

The theory of multi-layer networks has generated a very important set of tools to analyse the structure of networks of networks, where nodes can be different entities such as proteins and genes, and also interactions can represent different relations. However, we still need to develop multi-layer network tools specifically addressing the system biology challenges. A few works have already been published in this direction: in [2] multi-layer networks are used to investigate and combine a large set of gene expression data; in [3], multi-layer networks are used to identify cancer driver genes by combining transcription factor co-targeting, micro-RNA co-targeting, protein-interaction networks and gene co-expression networks; in [4] a comprehensive study of genotype and phenotype interactions of different diseases is launched. Moreover, system biology can be also enriched by integrating methods developed for multi-layer networks of the brain investigated both at the neuronal level [5] and at the level of brain regions [6].

12.2.2 Multi-layer Network Inference

Network inference is one of the major problems in systems biology as its main objective is to understand biological systems and the mutual influence of their elementary units. Theoretically significant progress has been made on multi-layer network inference methods. Several multi-layer network measures have been proposed, such as multilinks and multidegrees that are able to encode very significant information about the function of the multi-layer networks. These measures have been applied successfully for analysing multi-layer and multiplex network data (see Chapters 6 and 7). Additionally, several centrality measures capturing important multi-layer network properties have been proposed (see Chapters 6 and 7). However, the challenges posed by multi-layer network inference are far from completely solved. Specifically, a significant challenge is constituted by multi-layer network reconstruction based on the noisy evidence from different data types eventually including missing information. Moreover, also in the ideal case in which we assume we have complete information about a system (no missing data) and experimental evidence coming from many different data sets, an important open question is how to optimally represent the data in multi-layer networks. In particular, current research focuses on the problem of assessing the trade-off between having few layers and losing some information and having too many sparse layers each carrying a little information about the entire system.

Bayesian networks have been used during the last decades to successfully infer networks and integrate different data types. As such, they are a suitable framework for inferring multi-layer networks. However, this class of methods normally returns a sparse network with only important associations, and their inference processes are computationally intractable for large data. Another interesting class of methods that can be considered for inference of a multi-layer network is kernel-based methods, which provide a selective way of accounting for different data types, but a simultaneous modelling of different types of relations has not yet been developed with these methods.

Therefore, in the future it will be important to overcome the limitations of these methods and acquire a comprehensive network science and data science approach for inference of multi-layer networks which would have significant potential for the development of system biology.

12.2.3 Current Tools and Software Analysis and Visualization

Since the field of multi-layer networks is young, when it comes to software and visualization tools the options are increasing but still limited. Several researchers have open access subroutines published online (see GitHub); however, the number of developed software tools and packages available to analyse multi-layer networks is somewhat limited. Among these packages we mention Muxvis, which is a very useful platform to analysis and visualize multi-layer networks. MultiNets is another package developed in R, designed mostly for social networks. Multiplex and multigraph are another two R libraries for the analysis of multiplex networks and for visualizing multigraphs. MAMMULT is a collection of programs (in C and Python) for the analysis and modelling of multi-layer networks. Other repositories for multiplex network modelling, randomization and multi-layer network centrality methods can be found on GitHub page (https://github.com/ginestrab)

However, there is an urgent need to have more tools to deal with multi-layer networks and analysis, especially for interdisciplinary users. An example of such tools is described in Chapter 5, where information-theoretic tools are used to describe and classify monoplex networks. The main difference between such tools and those rooted in classical information such as Shannon entropy, is that they are invariant to changes of object description, and therefore this concept can be easily extended to cover multi-layer networks.

12.2.4   Dynamics of Multi-layer Networks

The investigation of dynamics on and of multi-layer networks is an important challenge for the field [1]. While diffusion has been extensively explored both on static multi-layer networks and on temporal networks, many questions related to the dynamics of molecular networks cannot be simply captured by a diffusion dynamics. In particular, in the context of temporal networks a central question is how to best partition the temporal windows for representing temporal networks as a sequence of static snapshots without losing information. Even more challenging are the scientific questions related to the description of temporal multi-layer networks or adaptive multi-layer networks that so far have not been extensively explored.

References

[1] Bianconi, G. *Multilayer Networks: Structure and Function*. Oxford: Oxford University Press, 2018.

[2] Li W, Liu CC, Zhang T, et al. Integrative analysis of many weighted co-expression networks using tensor computation. *PLoS Comput Biol*. 2011;7(6):e1001106.

[3] Cantini L, Medico E, Fortunato S, Caselle M. Detection of gene communities in multi-networks reveals cancer drivers. *Sci Rep*. 2015;5:17386.

[4] Halu A, De Domenico M, Arenas A, Sharma A. The multiplex network of human diseases. *NPJ Syst Biol Appl*. 2019;5(1):15.

[5] Bentley B, Branicky R, Barnes CL, et al. The multilayer connectome of *Caenorhabditis elegans*. *PLoS Comput Biol*. 2016;12(12):e1005283.

[6] Bassett DS, Wymbs NF, Porter MA, et al. Dynamic reconfiguration of human brain networks during learning. *PNAS*. 2011;108(18):7641–7646.

Index